The Discovery of Cosmic Voids

The large-scale structure of the Universe is dominated by vast voids with galaxies clustered in knots, sheets, and filaments, forming a great "cosmic web." In this personal account of the major astronomical developments leading to these discoveries, we learn from Laird A. Thompson, a key protagonist, how the first 3D maps of galaxies were created. Using nonmathematical language, he introduces the standard model of cosmology before explaining how and why ideas about cosmic voids evolved, referencing the original maps, reproduced here. His account tells of the competing teams of observers racing to publish their results, the theorists trying to build or update their models to explain them, and the subsequent large-scale survey efforts that continue to the present day. This is a well-documented account of the birth of a major pillar of modern cosmology and a useful case study of the trials surrounding how this scientific discovery became accepted.

LAIRD A. THOMPSON is Professor Emeritus of the University of Illinois at Urbana-Champaign. He has held appointments at the Kitt Peak National Observatory, at the University of Hawaii, and at the University of Nebraska, and is a member of the American Astronomical Society and the International Astronomical Union.

The Discovery of Cosmic Voids

LAIRD A. THOMPSON

University of Illinois, Urbana-Champaign

CAMBRIDGE
UNIVERSITY PRESS

University Printing House, Cambridge CB2 8BS, United Kingdom

One Liberty Plaza, 20th Floor, New York, NY 10006, USA

477 Williamstown Road, Port Melbourne, VIC 3207, Australia

314–321, 3rd Floor, Plot 3, Splendor Forum, Jasola District Centre, New Delhi – 110025, India

79 Anson Road, #06–04/06, Singapore 079906

Cambridge University Press is part of the University of Cambridge.

It furthers the University's mission by disseminating knowledge in the pursuit of education, learning, and research at the highest international levels of excellence.

www.cambridge.org
Information on this title: www.cambridge.org/9781108491136
DOI: 10.1017/9781108867504

First published 2021

Printed in the United Kingdom by TJ Books Limited, Padstow Cornwall

A catalogue record for this publication is available from the British Library.

ISBN 978-1-108-49113-6 Hardback

To my wife Jun and our children, Bolun and Bokei, the loves of my life, and to our cat Lily who behaves herself until we look away

Contents

Foreword

Laird A. Thompson and I entered the Astronomy program at the University of Arizona at the right time. The early 1960s had seen a burst of governmental funding for the sciences in general and for astronomy in particular, largely because of the space program. As a result of this funding and wise decisions made by the Department of Astronomy, headed by Professors Bart Bok, Ray Weymann, and then, after we graduated, Peter Strittmatter, new faculty, new telescopes, and new detectors were rapidly brought into the program and were in place by the time we arrived. Chief among these, for the purposes of the research described here in Laird's book, was the development of the Image Intensifying Tube (IIT). After the "white light" from a galaxy had been spread out into a spectrum by means of diffraction gratings, the spectral properties of the photons that carry information about the individual wavelengths emitted by the galaxies could be amplified without destroying the wavelength information. In this manner, one could record the spectrum of an individual galaxy in tens of minutes instead of the hours that were necessary just prior to our work. Thus, a "redshift" could be obtained for a galaxy and, using the Hubble-Lemaitre Law, its distance could be inferred. Laird and I recognized the potential of this device for placing the 3D knowledge of the distribution of galaxies on a much firmer footing.

What we found resulted in a profound change in the way humans look at the Universe. Instead of a field of randomly distributed galaxies with a few clusters of galaxies located here and there, the true nature was that galaxies are distributed along filamentary structures (which had already been called "superclusters"), and these filaments were found to be separated by huge, vaguely spherical, empty regions, which we called "cosmic voids" (a term that was entirely new). Clusters of galaxies occurred along the filaments, especially at places where multiple filaments intersected.

How big a deal is this change? In scientific terms, it represents the need for completely new physical processes that had not been previously considered, so it is, indeed, a big deal. For non-scientists, I find there to be a good analogy that can be made with nighttime images of the United States taken today and, say, 150 years ago. I envision a comparison between the 2D distribution of lights seen at night from orbiting spacecraft today with a hypothetical, similar map made long ago. In the older map, there would certainly be clusters of lights – cities with their streetlamps glowing. However, outside of the cities, there would be a roughly random distribution of lights coming from many small towns and perhaps a few farms with illumination. Since most of the population of the United States at that earlier time was rural, any kind of representation of the actual population distribution would show a largely random "field" of lights but with clusters colocated with the cities. In contrast, the current images we actually see of the nighttime United States, as detected from space, show very strong clustering and, importantly, the city clusters are strung out in a way closely analogous to superclusters. Obviously, one of the better examples is the megalopolis that stretches from Boston down to the Carolinas along the eastern seaboard. Interpretations of the different population distributions reveal profound changes in how Americans live their lives.

I want to make one more point about the work that Laird and I did. Personally, I think that a responsible scientist has (among others) the following two characteristics. First, the scientist needs to have a deep understanding of the foundations of the subject matter in question. Second, the scientist also needs to have deep skepticism about the standard picture that has arisen among subject-matter experts. Newton, in commenting on his new universal gravitation – which represented a huge advance in understanding – recognized both of these statements in his comment that he simply stood on the shoulders of giants in providing the rest of the world with the law of gravity. But in that era, he had no idea about what mechanism caused gravitation. In the case of the large-scale structure in the galaxy distribution, Laird and I worked very hard to understand what was known about how matter was distributed in the Universe. We talked about the opposing views of people like George Abell and Fritz Zwicky. We pored over their catalogs of data and wondered where their widely disparate overall viewpoints came from. We were not unique in realizing that the picture of that distribution rested on questionable assumptions. However, we were the first to demonstrate how shaky the basic concepts were at that time. We knocked down the ramparts that had been built on a bed of sand.

I am most grateful to Laird for having written this book. I have found it to be remarkably faithful to – at least in my own memories – what happened. We worked in a community of brilliant people and had guidance and encouragement from sources that were not always obvious to us. I am also grateful to all those colleagues, even the friendly rivals. Most of all, I want to express my appreciation to W. G. Tifft, who planted the first seeds of cosmic voids in our minds and then let us develop the field as we saw appropriate. He is a truly gifted scientist who is largely unappreciated. Finally, I note that in most of our joint papers, the authors are listed alphabetically. This places my name before Laird's in the author list. I want to be sure that readers understand that our work was a joint effort and that basically we deserve equal credit for our discoveries.

Steve Gregory

Preface

Modern cosmology rests on four solid pillars, each of which was assembled systematically from the meticulous work of observational astronomers during the past century. With this foundation in place, theoretical physicists and cosmologists have constructed a mathematical model of the Universe, precisely tuned to the extent that it has significant predictive power to infer conditions in the earliest epochs when our Universe first emerged from a hot, dense state. This is science at its best: the pinnacle of achievement for the cosmology community.

The four great pillars include the following.

The velocity–distance relation for galaxies reveals the expansion of the Universe. This relation was constructed from observations made between 1912 and 1927 and can be credited to Vesto Slipher, Edwin Hubble, and Georges Lemaître. A new feature was added in 1997 when a gentle *accelerated expansion* was detected based on the same velocity–distance concept but for faint and very distant supernovae. The supernova work was done by research groups at UC Berkeley and Harvard with standout performances by Saul Perlmutter, Adam Riess, and Brian Schmidt.

The lighter elements such as helium, deuterium, and lithium are synthesized during a three-minute hot and dense early phase of our evolving Universe. The connection between these element abundances and the physics of the early Universe was first recognized in the period 1946–1948 by George Gamow, Ralph Alpher, and Robert Herman (with additional key input from Chushiro Hayashi in 1950). The full impact came twenty to twenty-five years later when astronomers were in a position to place more stringent observational limits on the abundance of these elements.

The cosmic background radiation was first seen and identified for what it is, in 1965, by Arno Penzias, Robert Wilson, and Robert Dicke's cosmology group at

Princeton University. This was the "smoking gun" confirming that our Universe included an early hot and dense phase. The 1965 discovery meshed with and reaffirmed Alpher and Herman's work. Analyses of the cosmic background radiation became the richest source of information for the new field of precision cosmology.

The large-scale structure in the overall galaxy distribution sheds light on the process of galaxy formation and the spatial distribution of the still-mysterious dark matter. This is the most recently added pillar in cosmology and one that also provides a rich source of new research results. More than 60–70% of the volume of the Universe is occupied by cosmic voids with the remainder in a web of superclusters, filaments, and walls of galaxies defined by the dark matter. This book gives a firsthand account of the discovery of cosmic voids and the structures that surround them, all dating from its beginnings in 1978.

My primary collaborator in the discovery of cosmic voids and extended "bridges of galaxies" was Professor Stephen A. Gregory. After completing the first-ever wide-angle galaxy redshift survey, he and I mapped the 3D distribution of galaxies on a triangular plot that extended hundreds of millions of light-years into deep space, far beyond what had been done earlier. In our first map, we uncovered remarkably beautiful new features of the Universe. First among them were vast empty regions in the 3D distribution of galaxies, regions that we named voids. Next, we clearly detected the first bridge of galaxies connecting two rich galaxy clusters. This bridge is one component of what is now called the "cosmic web." Most significantly, our 1978 discovery spelled an end to the formal concept of "field galaxies," an idea that had been conjured in the imaginations of Edwin Hubble and Fritz Zwicky in the 1930s. "Field galaxies" were said to uniformly fill the Universe. Our 3D map showed nothing of the kind and turned the old view on its head. We knew the significance of our work, and we tried to make the most of our unique position. But as this book documents, things quickly went in directions we did not anticipate. On one hand, a fraction of traditional cosmologists labeled the newly identified large-scale structure in the galaxy distribution as unbelievable and ignored it. On the other hand, a number of those who did the follow-up work capitalized on the new discoveries by trying to attach their own names to it. In the period from 1986 to 2000, other astronomers were often given exclusive credit for what we had done. In this book, I aim to bring a little rational order to what happened and to make historical sense of the discovery process.

The construction of the fourth pillar of modern cosmology was an extended process that lasted nearly a decade. Similar processes occurred when the other pillars were revealed. This book traces the ins and outs, the foibles and successes of the discovery process. Our modest but epoch-changing redshift survey was followed by a series of further studies that confirmed our initial results and extended them, one step at a time, to include larger and larger volumes of the Universe. As the surveys grew larger, many additional cosmic voids were identified, and extended patterns were eventually detected in the galaxy distribution. The feature we identified as a bridge of galaxies connecting two rich clusters of galaxies turned out to be a small segment of what is now called the "Great Wall" of galaxies. By the turn of the twenty-first century, surveys more ambitious than ours showed many extended and interconnected structures. With forty years of hindsight, the process of discovery can be seen with better perspective and with greater clarity than when the events were happening.

For the convenience of the reader, each chapter of this book is more or less a self-contained unit, so the chapters need not be read sequentially. Chapter 1 starts with a description of the standard model of modern cosmology – providing a context as well as background information for the nonscientific reader – followed by Chapter 2 with a quick overview of the cosmic voids discovery story. Chapter 2 omits, however, many of the finer details. For those who are looking for a quick read and want to skip the numerous historical details sprinkled throughout the book, one might read Chapters 1 and 2 followed by Chapter 8. Chapter 8 summarizes several claims and counterclaims made during the discovery process by the scientists who were involved in the early observational research. Most importantly, in Chapter 8 I also present a timeline of the discovery of cosmic voids, placed side by side with the analogous timeline for the discovery of the cosmic microwave background radiation. The work on voids is in two columns, one for the observational work and a second for the theoretical models built to explain what we discovered in the galaxy distribution. The theoretical developments are traced step-by-step in Chapter 7.

The all-important discovery timeline is given in Table 8.2. Those who have previewed this book have repeatedly asked to have this timeline placed right upfront. They somehow think that the truth in defining who discovered the large-scale structure in the galaxy distribution will be resolved by studying this timeline. Independent of where this timeline is placed, obviously it can be accessed while reading any of the earlier chapters. More important, however, is that the finer details of the story described in the earlier chapters are needed to understand the fact that scientific discovery is an extended process that is punctuated by significant breakthroughs. This thesis was carefully explained in the 2013 book by S. J. Dick entitled *Discovery and Classification in Astronomy*. How else can anyone explain the behavior of leading researchers who refused to accept our early discovery, only to reverse direction later and begin to claim the discovery as their own?

Those interested exclusively in the story describing Gregory's and my pioneering work can read Chapters 1 and 2 followed by Chapter 5. Chapters 3 and 4 present a historical account documenting how astronomers from the 1930s through the 1970s were so fooled by Hubble's and Zwicky's assumption of a locally homogeneous Universe filled with field galaxies that they willingly ignored occasional evidence of the vast empty volumes that surround us. The book concludes in Chapter 9 with a discussion of vibrant ongoing cosmic void studies along with an up-to-date synopsis of the many ways cosmic voids are now being used in forefront investigations to test dark energy and to test models that contain modifications to gravity in the weak limit. The book flows from one chapter to the next, and I will be most pleased with those who read it from start to finish.

I make one technical point up front. Scientific discovery involves proving (or disproving) with sufficient evidence clearly stated hypotheses. This is the scientific method. My work with Gregory followed this format. There are some in the scientific community who lower their guard and confuse a "consistency argument" with a discovery. Someone might have a perfectly reasonable theory or hypothesis, but they may not have a clearly defined path to test it, or they lack the data to do so. From the outset of our work, Gregory and I defined a test for supercluster bridges and filaments in the local Universe – precursors to the cosmic web – and we systematically obtained the data to test for this structure. While testing this hypothesis, we stumbled across – and decisively detected – the vast empty voids that fill the Universe.

Consistency arguments often appear when speculative ideas are being actively pursued in science but before they are proven or disproven. Two examples will suffice. In 1755, the famous philosopher Immanuel Kant argued for the "Island Universe" theory to explain faint nebulous patches that had been seen scattered across the night sky. Kant conceptualized the idea that our Milky Way galaxy was one such system, 168 years before Hubble proved it to be true. Clearly, Hubble gets the scientific credit for the discovery: He provided the irrefutable evidence and not the speculative idea. The second example is a contemporary one. In cosmology today the phenomena associated with an inflationary phase of the early Universe are well known and are of great interest to many scientists. The theory of inflation suggests that the radius of curvature of the Universe zoomed up in size in an early epoch. Inflation seems capable of addressing fundamental assumptions used in the widely accepted standard model of cosmology (as Chapter 1 explains, the standard model of cosmology is called "LCDM"). While many astronomers and cosmologists find the theory of inflation compelling, it is not yet a proven theory. Steps that could make its case stronger – i.e., measuring twists in the polarized component of the cosmic background radiation – have been difficult to execute. The theory of inflation has not yet been proven or disproven with any clear-cut scientific test.

I was fortunate to have played a key role in defining and working on all of the earliest wide-angle galaxy redshift surveys: the Coma/A1367 supercluster, the Hercules supercluster, the Perseus supercluster, the A2179+A2199 supercluster, and the bridge of galaxies that links the Hercules supercluster with A2197 +A2199. I also contributed, in a minor way, to observations from the Arecibo 21-cm radio telescope to survey the Perseus region. My aim in writing this book is to review and to highlight the scientific significance of cosmic voids and the surrounding supercluster structure, and at the same time to share episodes of the discovery story of the cosmic web that have not been documented elsewhere.

Acknowledgments

Above all else, I thank my wife Jun Chen for her continued love and support and my two children, Bolun and Bokei, for bending and compromising while I indulged in writing a book that concerns issues that occurred long before my two children were born. I thank a number of colleagues who gave their time to check and correct my early drafts. First among these is, of course, Steve Gregory. I thank as well Leo P. Connolly who was a fellow graduate student at the University of Arizona, in the same class as me and Steve Gregory. J. Ward Moody – who worked with Robert Kirshner as a graduate student and with Steve Gregory as a postdoc – gave me structural advice on the opening chapters. Professor Moody continues to investigate void galaxies. Historian of science, Robert W. Smith checked my work in Chapter 3 concerning William Herschel, Henrietta Leavitt, Harlow Shapley, and Edwin Hubble. Professor Adrian Melott read and corrected my account of events in Chapter 7 on the development of the early theoretical models of galaxy formation. My trusted adaptive optics collaborator, Professor Scott W. Teare, advised me on key sections where a careful choice of words worked better than others. I thank Professor Volker Springel for his quick assistance in providing the highly relevant, as well as colorful, book cover image.

Acquiring permission to reproduce images was a challenge I had not anticipated, and I thank the following people for their help in this regard: Professor Andrew Fraknoi (for the image of George Abell), Professor Jaan Einasto (for images of himself and Mihkel Jõeveer), Dr. Thomas Fleming (for help with the Bok Symposium image), The Huntington Library Reference Services Manager Stephenie Arias (for images of Einstein, Hubble, and Humason), archivist Loma Karklins at Caltech (for images of Lemaître and Zwicky), archivist and librarian Lauren Admundson at Lowell Observatory (for the image of Slipher), *StarDate* Editor Rebecca Johnson at McDonald/UT Austin (for the image of de

Vaucouleurs), librarian Maria McEachern and archivist Lindsay Smith Zrull at the Center for Astrophysics (for images of Leavitt and Shapley), and Director of Strategic Communications Mary Beth Laychak at the Canada–France–Hawaii Telescope for handling copyright issues for the images of M15 and M51.

Thanks also to my astronomy colleagues who signed off on copyright permissions for quotes, graphs, individual photos, and group photos. These names I give without their esteemed titles in alphabetical order: Ken Croswell, J. Richard Gott, Martha Haynes, Fiona Hoyle, Mario Juric, Robert Kirshner, Adrian Melott, Jim Peebles, Sergei Shandarin, Stephen Shectman, Ravi Sheth, Paul Sutter, Massimo Tarenghi, William Tifft, Brent Tully, and Simon White.

Finally, I acknowledge the educational support I was fortunate to have received in both physics and astronomy at UCLA and in observational astronomy at the University of Arizona Steward Observatory. My late parents and other close relatives spurred my interest in science from an early age and supported me until this passion managed to take root.

It has been my pleasure to work with the professional editorial staff at Cambridge University Press, Vince Higgs and Henry Cockburn, and to have received advice from two anonymous referees who reviewed and supported my original book proposal.

While all of these individuals provided support, I am solely responsible for the contents of this book and any errors or misconceptions it contains. I welcome comments from informed readers so that any future edition(s) might be corrected and improved.

Abbreviations

a	Radius of curvature of the Universe
a(t)	Time-dependent radius of curvature of the Universe
CDM	Cold dark matter
CfA1	First Center for Astrophysics redshift survey (initiated by M. Davis)
CfA2	Second Center for Astrophysics redshift survey (initiated by J. Huchra and M. Geller)
CMB	Cosmic microwave background radiation, a remnant from the origin of the Universe
DEFW	CDM model of structure formation by Davis, Efstatiou, Frenk, and White (1985)
WFDE	CDM model of structure formation by White, Frenk, Davis, and Efstatiou (1987)
KPNO	Kitt Peak National Observatory
LMC	Large Magellanic Cloud – a satellite galaxy in orbit around our Milky Way
Mpc	Megaparsec, a unit of distance where 1 Mpc = 3.26 million lightyears
SMC	Small Magellanic Cloud – a satellite galaxy in orbit around our Milky Way
SDSS	Sloan Digital Sky Survey with Data Releases abbreviated DR9 (for the ninth)
z	Galaxy redshift = observed wavelength/standard reference wavelength on Earth
ZA	Zeldovich Approximation: a mathematical formulation of starting conditions for galaxy formation models

1

Understanding the Foundations of Modern Cosmology

Over the last 100 years, theoretical physicists and observational astronomers have uncovered the birth story of our Universe and have coaxed its key physical properties from observations of the sky. The first steps were taken by a handful of great scientists early in the twentieth century: Albert Einstein, Alexander Friedmann, Vesto Slipher, Abbé Georges Lemaître, Edwin Hubble, and George Gamow. These pioneers and others who followed in their footsteps were able to peek behind a curtain that has now been flung wide open. An evolving model of our expanding Universe has taken center stage, and its characteristics are nothing short of breathtaking. Its current and most popular form is called the "LCDM" model. It begins in a state where all regions of space are nearly uniformly filled with an unimaginably hot and high density of energy.

The LCDM model is a sophisticated and refined hybrid of the Big Bang theory that was sketched in its most rudimentary form by Georges Lemaître around 1930. Lemaître was the first to hypothesize that the Universe began in a high-density state with a tiny "radius of curvature," only to evolve into our current state with a huge "radius of curvature." The mathematical basis for his model – as well as the basis for the LCDM model – derives from Einstein's theory of general relativity in a form suggested by Alexander Friedmann.

The name LCDM was selected to highlight two key constituents of the Universe that were not part of Lemaître's original concepts. The "L" stands for Lambda, an antigravity force that is also called the "cosmological constant." It was introduced by Einstein in his original model of the Universe. Lambda was only occasionally employed in Big Bang models in the 1950s and 1960s, but in today's LCDM model, Lambda accounts for 69% of the mass–energy content of our present-day Universe. The second primary constituent is cold dark matter (CDM). CDM was first conceptualized in 1978. Although it played a key role in the discovery of cosmic voids, the exact underlying nature of CDM is still a mystery, in the sense that no one knows the composition of this elusive constituent. The fact that it is called "cold" means that it moves around under the force of gravity at relatively low speeds (i.e., it is likely to have a particle-like nature unrelated to electromagnetic radiation that moves at the speed of light). According to the LCDM model, each galaxy forms in the central region of an extended "halo" of CDM. These halos begin to form first and then begin to settle into a 3D filamentary web that forms the scaffolding for galaxy formation. In our current epoch, there are huge empty regions – cosmic voids – located between the sheets and filaments of the 3D web of dark matter. Of course, these vast empty regions are the central focus of this book. CDM accounts for 26% of the mass–energy content of the Universe. What remains is a mere 5%. This is our fraction: the material in us and in the stars and planets that reside in galaxies around us. We can only see the 5%; however, it is a tracer that lights up the dark matter halos that are situated in the cosmic web and also allows us to detect the outward motion caused by the expansion of space, as well as the added accelerating push of Lambda.[1]

1.1   Predictions of the LCDM Model

The evolutionary LCDM model links the physical nature of matter and energy at the starting point of our Universe to the features of the cosmos that we see today. Emerging from the earliest exotic phase of high-energy phenomena is a Universe uniformly filled with and dominated by extremely hot high-energy electromagnetic waves. In standard jargon, these are light waves (alternatively called "photons") that possess the highest possible energies. These so-called gamma rays lose energy and therefore cool as the space that contains them expands. As the temperature begins to drop, various constituents freeze out of the background energy field of gamma rays – quarks with their associated gluons, neutrons, protons, electrons, neutrinos – one component at a time. Here is how the freeze-out occurs. Pairs of gamma rays are capable of spontaneously generating particle pairs: a proton and an antiproton, a neutron and an antineutron, or an electron and an antielectron (also called a "positron"). In a theoretical sense, the antiparticle is a "mirror image" of the actual particle. Each particle pair is associated with a specific gamma-ray energy, namely the energy equal to the total rest mass $E = mc^2$ of the particle pair. Once the decreasing gamma-ray energy in the evolving Universe drops significantly below the $E = mc^2$ limit required to produce a specific particle pair (a decrease in energy caused by the expansion of space as the radius of curvature of the Universe grows larger and larger), no more particle pairs of that type are created. After the freeze-out for each specific particle in question, nearly all particles eventually meet a corresponding antiparticle and annihilate, and the energy from the pair production goes back into gamma rays. By somewhat of a magical quirk of nature called "CP violation," there is a very slight imbalance in the creation of particles and antiparticles, so after the annihilation of particle pairs is complete, a tiny fraction (about one part in a billion) of the initial energy remains in the remaining mass of the regular particles that have frozen out. The remainder of the energy is redeposited back into the bath of cooling gamma rays. All antiparticles are lost in the process. Figure 1.1 sketches one point in this early phase.

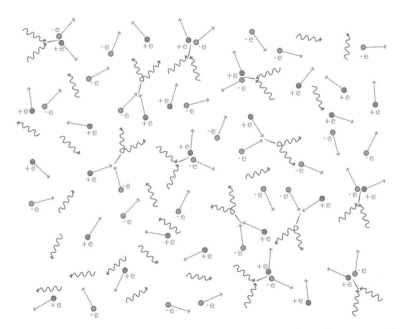

Figure 1.1 Electron-positron pair production. This schematic represents the early Universe where electron and positron pairs are spontaneously produced by high-energy gamma rays as well as the reverse: an electron and positron pair annihilates to produce gamma rays. Sine waves with arrows represent the trajectories of gamma rays, and straight lines with arrows represent electrons (−) or positrons (+). The threshold temperature for $e^- - e^+$ pair production is ~10 billion K, and this temperature occurs when the Universe is several seconds old as the Universe expands and cools. By this time, more massive particles like quarks, neutrons, and protons have already "frozen out" of the expanding plasma in a process similar to that shown here, but at higher temperatures and therefore higher energies. Image copyright A. Bokei Thompson.

After the particles have frozen out of the expanding plasma, the first large structures begin to take shape. These consist of extended diaphanous irregularities of CDM, as it begins to gather itself together under the force of gravity to eventually define seed structures that ultimately trigger the formation of the first stars and galaxies. CDM starts to congregate even when our 5% fraction – the electrons, protons, neutrons – remains uniformly distributed. Only when the temperature cools further and the electrons join the atomic nuclei does the ordinary matter detach itself from the still-brilliant background gamma-ray radiation. Once detached, the matter is set free to fall into the CDM structures to begin the creation of stars and, ultimately, galaxies in the CDM halos. The 69% contributed by Lambda has little effect early on, but now that the Universe has expanded to its current radius, Lambda has begun to dominate gravitationally. In the future, the effects of Lambda are predicted to overpower gravitational forces on the largest scales. The presence of Lambda was discovered in the late 1990s when astronomers organized surveys to identify and measure the brightness of very faint and distant supernova explosions designated "SN Ia." These explosions just happen to display a very specific maximum brightness and are therefore one type of "standard candle" that astronomers have been fortunate to discover. As these distant SN Ia appear somewhat fainter than what is expected in a simple expanding Big Bang model today, they reveal the existence of Lambda.[2]

The searing hot and extremely dense phase identified with the origin of the LCDM model was just as internally infinite in 3D space as our current Universe is infinite. Einstein's field equations of general relativity are used to calculate the dynamic expansion of this infinite manifold, including the outward accelerating effects that are caused today by Lambda. General relativity involves a set of ten field equations that are based on the ordinary three dimensions of space (x, y, z) plus a time dimension that is written as ct where c is the speed of light and t the time. According to the most recent measurements, our (x, y, z, ct) manifold is extremely close to being flat on the largest scales; in other words, space appears to be "Euclidean." The four-component (x, y, z, ct) manifold is the underlying basis for the theory of general relativity introduced in 1915 by Albert Einstein (1879–1955). It involves an exact mathematical balance between the mass–energy density at any given point in the Universe and the curvature of space at that same point.[3] It is from the mathematical solution of Einstein's field equations of general relativity that the radius of curvature of the Universe emerges.

Albert Einstein (1879–1955)

Albert Einstein was born in Ulm in southern Germany, close to both Munich and Zurich, Switzerland. He was educated in a Munich gymnasium, and at age seventeen, enrolled in what is now ETH in Zurich, where one of his classmates was Marcel Grossmann. When Einstein could not find a teaching position upon graduating from ETH, Grossmann's father found for Einstein a position in a federal patent office in Bern, Switzerland, in 1902. This position gave Einstein sufficient free time after office hours to write papers in theoretical physics. He quit the patent office in 1909, at age thirty, when he received his first university teaching position. By then, Einstein was widely acclaimed for a series of papers published in 1905, including his theory of special relativity. In this period, Grossmann had become a professor at ETH and an expert in Riemannian geometry, a fact that aided Einstein in his development of general relativity theory. Einstein contributed to many fields of physics and is considered to be one of the most brilliant scientists of all time. By 1917, Einstein published his static and closed model of the Universe, a model that was largely motivated by his interest in principles espoused by physicist and philosopher Ernst Mach regarding the nature of inertia and inertial rest frames. Modern physical cosmology got its start with Einstein's 1917 model, but it quickly developed on its own with contributions by de Sitter, Friedmann, Lemaître, and others. In 1933, Einstein and his second wife Elsa emigrated to the United States from Germany and took up residence at the Institute for Advanced Study in Princeton, New Jersey. After interactions with astronomers like Hubble in the early 1930s, Einstein quit working in cosmology by 1934. In 1954, Einstein abandoned his former interest in the work of Ernst Mach. Primary source: Pais (1982). Image courtesy of the Observatories of the Carnegie Institution for Science Collection at the Huntington Library, San Marino, California.

The solution to Einstein's field equations that is universally accepted today results in an "ever-expanding" manifold. This simply means the following. In the solution to Einstein's equations of space-time, there is a scale factor that is designated sometimes with the letter "R," and at other times with the letter "a." Here I will use "a." While this scale factor could be constant, any model of the Universe with a constant "a" is unstable: With the slightest disturbance, the manifold will either expand or contract. Once it is destabilized, "a" becomes time-dependent, so it is written as "$a(t)$." In Einstein's very first model of the

Universe, he assumed that space was positively curved and static (Einstein 1917). In positively curved space, Euclidean geometry does not apply. Being static meant that the scale factor $a(t)$ was a fixed constant. Einstein stabilized his model and forced it to be static by choosing a value for Lambda (the cosmological constant) to precisely counterbalance the positive attraction of gravity of all the matter in his Universe. Even though Einstein's first model was philosophically beautiful, it quickly fell by the wayside because, as noted earlier, the solution was unstable and therefore not applicable to our Universe. In 1922, Russian scientist and mathematician Alexander Friedmann (1888–1925) found alternate solutions to Einstein's field equations of general relativity that allowed the scale factor $a(t)$ to change with time (Friedmann 1922).

Unknown to Einstein and Friedmann, American astronomer Vesto Slipher (1875–1969) had, by that time, already seen hints that space is expanding. During a nine-year span (1912–21), Slipher (1917) obtained the first high-quality galaxy spectra from which he was able to measure galaxy velocities based on the Doppler shift.[4] Individual stars had already been measured by astronomers to have Doppler shifts in the range of 1 km s^{-1} to 5 km s^{-1}. Some star clusters (component parts of the Milky Way) had Doppler shifts up to 60 km s^{-1}. However, Slipher found the Doppler velocities of spiral nebulae to be hundreds of km s^{-1} and some exceeded 1000 km s^{-1}. No one could explain it! However, the astute British astrophysicist Arthur Eddington included a table of Slipher's Doppler shifts in his 1923 book *The Mathematical Theory of Relativity* (Eddington 1923, p. 162). It was a portent for the future.

Eventually, it was Slipher's and Edwin Hubble's observations, Einstein's general relativity, and Friedmann's mathematical contributions that became the building blocks used in the period 1927–33 by Abbé Georges Lemaître (1894–1966), a priest and physics professor from Belgium, to sketch the outlines of an evolving model explaining the origin and subsequent evolution of our Universe. The Universe is said to be expanding because the radius of curvature, $a(t)$, is increasing with time.[5] Lemaître identified the birth of our Universe – the era when the radius of curvature, $a(t)$, closely approached zero – with what he called the "Primeval Atom" or the "Cosmic Egg." The Universe, he suggested, started as a compact seed that expanded to become the Universe we see today. By the time Lemaître's model faced its first viable competitor – the Steady-State Theory – George Gamow (1904–68) had shown that the early Universe was hot, and Lemaître's model was renamed the Big Bang. The alternative Steady-State Model (Bondi and Gold 1948; Hoyle 1948) included an expanding Universe with the continuous creation of matter, thereby avoiding the early high-density phase. During the period 1950–1965, there was an open debate as to the merits of these two models: Steady State versus the Big Bang. While Lemaître had the general idea correct – that the very early Universe was extremely dense and resembled, in some ways, an atomic nucleus of infinite extent – he made a few incorrect assumptions. First, he identified what we call cosmic rays (high-velocity atomic nuclei that bombard the Earth from outer space) as remnants of the Big Bang. Today, we know that cosmic rays have no direct connection to the Big Bang. Second, he did not specify whether the early Universe was hot or cold. What Lemaître certainly overlooked was the importance of high-energy electromagnetic waves – specifically gamma rays – that were dominant in the earliest phases of the Universe.

Vesto M. Slipher (1875–1969)

Born on a farm in central Indiana, V.M. Slipher graduated from Indiana University, Bloomington, in June 1901, with specialties in mechanics and astronomy. Based on the recommendation of his faculty advisor, he was eventually hired by Lowell Observatory in Flagstaff, Arizona, where he was given the job of commissioning the observatory's new state-of-the-art spectrograph. To establish the wavelength calibration for the spectrograph, Slipher made his own arc lamps and attached them to high-voltage Leyden jar capacitors that he built and charged up in advance. After proving to himself and to Lowell that he could successfully detect spectra of stars and planets, in 1909 he turned to the more difficult problem of recording spectra of spiral nebulae. Director Percival Lowell thought spiral nebulae would be ideal targets because they could be sites of star and planet formation. By 1912, Slipher had successfully measured the first Doppler velocity of a galaxy (the Andromeda galaxy), and by 1923 he had measured forty others. Slipher showed that the spectrograph's efficiency when detecting spiral nebulae was dependent on having a wide entrance slit and a short focal length camera lens in the spectrograph. It was not at all dependent on the aperture of the telescope he was using: Slipher used the Lowell Observatory 24-inch refractor for all of his measurements (as discussed by Thompson 2013). Slipher remained at Lowell his entire career and became Observatory Director. In the 1930s, he oversaw an extensive survey program aimed at detecting planets. It was in this survey that Clyde Tombaugh, under Slipher's direction, discovered the dwarf planet Pluto. Slipher resigned as Lowell Observatory's Director in 1952 at age seventy-seven. Slipher was a modest man despite his extraordinarily successful career in astronomy. Primary source: Hoyt (1980). Photo reproduced with permission: Lowell Observatory Archives.

Based on the most recent results from LCDM, the Universe began when the scale factor $a(t)$ was at its minimum, some 13.81 billion years ago. At that time, 3D space was bursting with gamma rays. As the scale factor of the Universe $a(t)$ increased from an early near-infinitesimal value, the temperature and the density dropped in a mathematically predictable manner. A similar effective decrease in temperature can be simulated in a laboratory here, on Earth, by building an oven with sliding walls. If the walls of the oven are moved slowly outward, the oven's volume increases, and the oven temperature drops. Richard Tolman (1881–1948) proved mathematically in 1931 that when the Universe expands slowly, the radiation it contains loses its energy ever so slowly, but it maintains its original energy (spectral) distribution (Tolman 1931).

Initially the temperature was so high that atomic nuclei were unable to hold onto electrons to form the neutral atoms we know today, because ubiquitous gamma rays from the early Big Bang would immediately kick the electrons free. But as the radius of the Universe continued to grow and the temperature continued to drop, eventually, the Universe cooled sufficiently for electrons to remain in orbit around atomic nuclei, thereby making ordinary neutral atoms possible for the first time. Observations of the sky interpreted with the help of the LCDM model show us that this momentous event – when electrons began to orbit atomic nuclei – happened when the Universe was 378,000 years old (as measured from the time of the initial hot and dense beginning). In our current epoch, $a(t)$ continues to increase and the temperature and average density continue to drop. All electrons, protons, and neutrons in the Universe, including those in our bodies, went through these early transitory states. An army of astronomers and physicists busy themselves by checking that the evolutionary predictions of LCDM fit the most up-to-date measurements obtained from the sky. This is the current state of affairs in cosmology.

1.2   Nucleosynthesis and the Cosmic Background Radiation

It was in the mid-1940s when Gamow, a Soviet émigré to the United States and professor at George Washington University, began to correct key shortcomings of Lemaître's model. This was the era of both atomic and hydrogen bombs when physicists were thinking hard about nuclear reactions, and Gamow and his students tried to figure out whether there could have been nuclear transformations in a hot early phase of the Universe that were responsible for the relative abundances of the elements that we see around us: hydrogen, deuterium, helium, lithium, beryllium, and so on.[6]

Today, astronomers find our Universe has a uniform composition with, on average, a density of only one atom per cubic meter, consisting primarily of hydrogen (74% by mass) and helium (25% by mass), with all other elements contributing to the tiny remainder. In an early phase of his work, Gamow gave to his research student Ralph Alpher (1921–2007) the exercise of calculating what happened to the nuclear particles in the hot, dense early Universe, and in 1948 Alpher was the lead author on the important paper by Alpher, Bethe, and Gamow (1948) entitled "Origin of the Chemical Elements." This paper reports the astounding result that naturally occurring nuclear fusion processes in the hot early Universe can explain the origin of some atomic nuclei when the initial conditions provide only protons (hydrogen nuclei) as a starting point. It was the first baby step into the field of cosmological astrophysics, now called "nucleosynthesis," which showed how the atoms around us – and from which we are made – have a direct connection to an evolutionary model of the Universe. It would not be until later that Hoyle and Tayler (1964) helped by extending the picture with a reliable calculation of the predicted abundances of hydrogen and helium with a more comprehensive summary by Wagoner, Fowler, and Hoyle (1967).

Late in 1948, Ralph Alpher teamed with Robert Herman (1914–1997) and introduced into their nuclear physics calculations the dynamic expansion of the Universe with its monotonically decreasing temperature and density. By doing so, their calculations – and those of Gamow – became more realistic, but more importantly, it led them to consider the fate of the high-energy gamma rays involved in the early nuclear transformations. Gamow (1948a, 1948b) with Alpher and Herman (1948) predicted that today we should see not gamma rays but microwaves coming from all directions in the sky, and these microwaves should have a characteristic temperature of 5 K. Detecting this background radiation could provide direct evidence of the Hot Big Bang. In 1948, when this 5 K background radiation was first discussed by Gamow and his students, radio receivers capable of detecting it were not available, so no further work was done at that time.

When the remnant radiation from the early Universe was finally discovered in 1964, it was found somewhat by accident. Arno Penzias (b. 1933) and Robert Wilson (b. 1936) had been hired by Bell Laboratories in Holmdel, New Jersey, to operate and observe the sky with a microwave radio antenna originally built for the Echo satellite experiment. They worked extremely hard to understand all sources of background noise in their radio antenna, noise that might be a problem whenever the antenna received signals from the sky. But they were unable to remove all background noise. What they first ascribed to noise in their radio antenna and receiver system, they soon realized, was actually a nearly uniform microwave "hum" emanating from all directions in the sky. At that time Penzias and Wilson were totally unaware of Alpher and Herman's work that had been completed 17 years earlier. But at nearby Princeton University, another group of physicists under the direction of Robert Dicke (1916–97) had redone the same cosmology problem that was first solved by Alpher and Herman, and they were aiming to detect the cosmic remnant radiation, too. Once the two groups, one from Bell Labs and the other from the Princeton Physics Department, exchanged notes, they published back-to-back papers in the *Astrophysical Journal Letters* in 1965 describing the discovery, and a new era of cosmology was born.

Georges Lemaître (1894–1966)

Abbé Georges Lemaître was a Jesuit priest from Belgium who was also an honored veteran of World War I. He received his first PhD degree in 1920 from the Catholic University of Leuven in mathematics and physics, at which point he won a scholarship from the Belgium government that allowed him to study abroad for two years. In 1923, he left for one year in Cambridge, England, and for a second year at MIT. At MIT, he earned a second PhD degree. Before returning home in 1925, he traveled by train to the western US to visit Hubble at Mt. Wilson and Slipher at Lowell Observatory. In 1927, after returning to Europe, he used his theoretical expertise along with Hubble's and Slipher's observations to derive the first relativistic model of the expanding Universe. The resulting research paper was translated into English and published as Lemaître (1931a). His original model took Einstein's static model of the Universe as its initial state, but by 1930 Lemaître suggested instead that the starting point was a "Primeval Atom" or "Cosmic Egg." Lemaître associated the beginning of time with the point when the entropy of the system (its randomness) began to grow. These ideas are the origin of what, today, we call the Big Bang. In 1946, Lemaître published in French a book that was translated into English in 1950 entitled "The Primeval Atom: A Hypothesis of the Origin of the Universe." Primary source: R. Berendzen, R. Hart & D. Seeley (1984). Photo by permission: Caltech Image Archives.

Penzias and Wilson's uniform microwave "hum" from the sky was soon confirmed by other physicists (two of these confirmations actually pre-dated the detection in Holmdel), and the temperature of the background radiation was found to be in the range of 3 K, not too different from Alpher and Herman's predicted 5 K. The hot Big Bang model thereby won a monumental victory over the Steady State model because there was no straightforward way for this background radiation to be explained by Steady State proponents. The radiation was named the Cosmic Microwave Background (CMB). Given the significance and wide acclaim ascribed to this discovery, it is no surprise that Penzias and Wilson were awarded the highest accolades available in physics and astronomy.

In the 50 years following Penzias and Wilson's work, ~100 separate experiments have further probed and analyzed the CMB radiation for evidence relating to the origin of our Universe. NASA's COBE (Cosmic Background Explorer) was a satellite launched in 1987. It made some of the first giant steps beyond the original discovery and proved to an incredible level of precision that the CMB radiation follows, indeed, a nearly perfect thermal distribution that is called in physics a black body spectrum. This is precisely what a Hot Big Bang model predicts. Microwave detector technology has steadily improved even further since 1990, and the most successful CMB experiments have been conducted in the last 20 years. Those with the greatest impact include BOOMERanG, DASI, WMAP, and Planck. BOOMERanG was an experimental package tethered to a high-altitude weather balloon and lifted to the edge of space; DASI was built and operated on a telescope located at the South Pole; WMAP was a satellite boosted into space by NASA in 2001; and Planck was a satellite that was sent into space by NASA's European counterpart, ESA, in 2009. The final Planck results describing the nature of our Universe became available in 2015. Each experiment improved the CMB detection and built on the earlier results. What we have learned from these experiments is astounding!

The microwave sky-background "hum" is very smooth – smooth with tiny fluctuations of only ~20 parts per million over a scale of 10 arc minutes (1/3 the Moon's diameter) – but these tiny irregularities hold a wealth of information. It is from a precise astrophysical analysis of these tiny CMB fluctuations, along with other astronomical constraints, that many key quantitative measures of our Universe have been deduced. For example, the analysis of the Planck satellite results has shown that it has been 13.81 billion years since the Universe began to expand from its hot and dense early phase. The current average density of the Universe appears to be extremely close to (if not identical to) the density required to make the geometry of space Euclidean. As mentioned above, the mass–energy content of our Universe consists of 69% antigravity force (designated above as Lambda) that some have associated with dark energy, 26% the mysterious dark matter, and 5% ordinary atomic particles: electrons, protons, neutrons, etc.

Whenever observations of the CMB irregularities are fit to the LCDM model, cosmologists make several simplifying assumptions. First, they assume that we live in an expanding Universe described by Einstein's equations of general relativity. A specific set of equations called "Friedmann-Lemaître-Robertson-Walker" (FLRW) is used, and they are applied to a space-time manifold of infinite spatial extent. Whether our Universe is actually infinite is a question that astronomers may never resolve because observations of the most distant realms are limited by the light travel time corresponding to the age of the Universe. We can see any object that emitted radiation towards us 13.81 billion years ago (i.e., the age of the Universe). We know that telescopes can probe only within a spherical volume with a diameter slightly less than twice that number. This defines our spatial horizon, and we have little to no direct knowledge beyond what we can see. Then, there are two separate infinities associated with time: into the future and from the past. Although the period of maximum temperature and density associated with the Big Bang happened at a time in the past that LCDM calls the "origin" of our Universe, there might be a longer and complex past history to the Universe. Some cosmologists have searched microwave maps of the CMB, looking for remnant signals that may have been imprinted on the CMB signal from an epoch that preceded our Big Bang era. Others suggest that the Universe might be oscillatory with recurring Big Bangs. The other assumption of cosmologists is that a future infinity is associated with the expansion of our Universe: into the infinite future. But just like the spatial infinity beyond our horizon, the assumption of an infinite future appears to be unavailable for us to investigate.

The LCDM model is now being extended to include the concept of inflation, a hypothesized period in the very early Universe when the value of scale factor $a(t)$ zoomed up from an initially tiny value to a new value at least 10^{30} times larger than when inflation began. While at first this theoretical concept seemed far-fetched, it has gained traction for several reasons. First, observations of the CMB radiation indicate that the CMB temperature is highly uniform over the entire sky. Say, we look at the CMB in two opposite directions and compare the results. The thermal properties appear to be essentially identical in the two far-separated regions. But one might ask: how can this happen? These two regions of the Universe, both at an equal distance from us but in opposite directions, can never have been in direct physical contact if we accept conventional LCDM cosmology. However, the theory of inflation allows these distant regions to have been in close contact with each other before the inflationary expansion phase began. Second, inflation is a way to solve the apparent mystery that our Universe appears to be generally homogeneous on the largest scales. Under the influence of inflation, all early irregularities are greatly diminished in their amplitude. Fortunately, an observational test to measure the polarization of the CMB radiation could shed some light on whether inflation did occur in the early Universe. The answer may be known within ~10 years, but already many cosmologists are assuming that inflation is next to necessary: it has not been proven, but it is consistent with what we know today.

1.3   Hubble Finds a Homogeneous Expanding Universe

For the Universe to be homogeneous simply means that the contents of one particular volume is similar to that of any other volume as tested within the entire visible realm. To be isotropic means that no direction is a preferred direction. For example, astronomers have never identified any particular place in outer space that looks significantly different than any other place. There is no single "central" object nor any group of "central" objects in the Universe.

Homogeneity and isotropy are assumptions that have been made repeatedly by cosmologists, independently of what they knew about the Universe. Perhaps the best example is when Einstein made his first model of the Universe. It was based on general relativity, of course, and he assumed homogeneity and isotropy. But in 1917, no one knew for sure about the true nature of external galaxies, and Einstein's "Universe" contained only stars. Einstein was not a student of astronomy, even though he would later meet with and discuss the nature of the Universe with many prominent astronomers, including Wilhelm de Sitter and Arthur Eddington in the 1920s and with Edwin Hubble and other California astronomers in the 1930s.

To determine if the Universe is homogeneous and isotropic, astronomers have traditionally relied on the galaxy distribution and have studied its properties across the sky. This effort started in the late 1700s, with a monumental visual search for faint nebulae by Sir William Herschel and his sister Caroline in England. William Herschel's son John Herschel completed the survey by including the southern sky as viewed from South Africa. In this first search to identify faint nebulous objects, the Herschels described and catalogued thousands of objects by staring through the eyepiece of their telescope. In this regard, they were the first scientists to see galaxies deep into the Universe. William Herschel noted and openly discussed the fact that faint nebulae are not randomly distributed on the plane of the sky, and he stated that they appear, instead, to be grouped together in "strata", long linear structures that stretch many degrees across the sky. This amazing early result and several subsequent investigations by scientists like Shapley and Hubble are carefully documented in Chapter 3. In these early surveys, the analysis and conclusions were restricted to 2D: galaxies were seen projected onto what we see as the spherical surface of the sky.

By the mid-1920s, Edwin Hubble (1889–1953) had identified our proper place in the Universe relative to external galaxies. He did so by first determining distances to the closest neighboring galaxies like the Andromeda nebula, and in the 1930s, he extended our knowledge of the Universe far beyond what his predecessors had accomplished. By 1936, Hubble had completed the first modern test for homogeneity and isotropy by counting faint galaxies across the sky. He concluded that the distant galaxy sample indicates homogeneity and isotropy on the largest scales. That part Hubble got right. He figuratively stumbled, however, when he suggested that on somewhat smaller scales, galaxies are spread through space in a way that is also homogeneous and isotropic. Hubble made the latter claims, despite reports from his contemporary and adversary Harlow Shapley (1885–1972) that in the nearby Universe the galaxy distribution is far from uniform. While Shapley was right, Edwin Hubble's other work was so highly regarded and influential that a majority of astronomers adopted Hubble's view of the galaxy distribution.

In the 1930s, Shapley and his assistant Adelaide Ames began a photographic investigation of the brightest ~1,300 galaxies in our vicinity of the Universe. In a general sense, they were repeating the work of Herschel, but they did so with photographs rather than with the naked eye. The Shapley and Ames study revealed a significant asymmetry in the galaxy distribution, showing that there are many more bright galaxies in the north galactic hemisphere relative to those in the south galactic hemisphere. This asymmetry, along with the "strata" seen by Herschel, provided the first evidence for an inhomogeneous local galaxy distribution.

Another great step forward in cosmology was made in the early 1930s, with the decisive confirmation of the velocity–distance relationship for galaxies by Hubble and his self-taught assistant Milton Humason. This profound cosmological result was first revealed in an obscure paper by Lemaître in 1927. Lemaître used galaxy velocities measured by Vesto Slipher from spectra taken 10 years earlier at Lowell Observatory in Flagstaff, Arizona, and distances determined by Edwin Hubble. Lemaître's velocity–distance relationship was clear, but it took subsequent observations by Hubble and Humason at the 100-inch telescope at Mt. Wilson Observatory in California to drive the point home.[7] Despite Humason's somewhat simple-minded demeanor, he was an assiduous observer who, like Slipher before him, collected spectroscopic exposures that were many hours long. In the most extreme cases, for the faintest galaxies, a single exposure might span several nights: at the end of the first night, the shutter on the photographic plate holder would be closed and the plate holder stored in a darkroom during daylight hours, to be reopened again only when darkness returned the following night, at which time the exposure was resumed. The velocity–distance relation is one of the cornerstones of modern cosmology because it provides decisive evidence for the expansion of the Universe. Figure1.2 displays Hubble and Humason's velocity–distance results from 1931.

It is unclear whether or not astronomers in the 1930s realized that the velocity–distance relationship could be used as a tool to reveal a 3D map of the galaxy distribution. This is the key to the discoveries Gregory and I made beginning in 1978. The concept is simple: for every galaxy with a measured velocity, the Hubble-Humason relation can be used to obtain an inferred distance. Thus, a 2D map of galaxies (as projected on the sky) plus the derived distances for all galaxies in the sample yields the 3D map. The problem in the 1930s was a practical one: obtaining nicely exposed spectra for any reasonably large sample of galaxies was not feasible. Each spectrum has to be detailed enough to reveal subtle emission and absorption features in the light recorded from the galaxy. Once they are recorded, these features are compared with their known rest-frame wavelengths (as seen in atoms in a lab on Earth). The comparison between our rest-frame and that of the distant galaxy yields the cosmological expansion velocity. But by the mid-1950s, a total of only ~800 galaxies had measurable spectra, so anyone who dreamed of a 3D analysis of galaxies would see that such a study was impractical in that era. I have seen no suggestion of it in print. Only with 3D information can 3D structures be detected.

The first attempt to study the 3D distribution of galaxies came in the 1950s. As described in Chapter 4, Gerard de Vaucouleurs tried to understand the nature of our Local Supercluster by identifying galaxy groups and galaxy "clouds" (extended and loose collections of galaxies) from his own galaxy catalogue, a catalogue that was based on the Shapley-Ames bright galaxy sample. Instead of determining distances to individual galaxies as described above, he estimated distances to galaxy groups and "clouds," each treated as a collective unit, and from this he created a 3D map showing the results. Given the limited sample of relatively local galaxies with measured velocities available in de Vaucouleurs' day (a sample of ~800 galaxies), he identified many structures but was unable to discern any meaningful results for the volume of space beyond the Local Supercluster.

The work of one generation inevitably passes on to the next. Hubble died in 1953, but his rival and contemporary Shapley lived on for another 20 years. Shapley must have felt gratified to see the results of two massive (nearly) all-sky photographic surveys that were completed in the 1950s. Both corrected the main deficiency in Shapley's work: his telescopes produced photographs that were not uniformly sensitive across his photographic field of view, thus preventing him from precisely characterizing

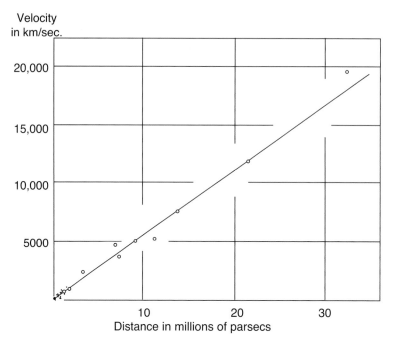

Figure 1.2 Hubble-Humason velocity-distance relation. This historic diagram decisively demonstrates a linear relationship between a galaxy's apparent velocity and its distance – the higher the apparent velocity, the greater the distance – providing clear evidence for the expansion of the Universe. The small solid points in the lower left corner show Hubble's 1929 results for single galaxies, and the open circles extending to the upper right show Hubble and Humason's 1931 results for clusters of galaxies. Modern-day recalibrations of the velocity–distance relation reveal that Hubble systematically underestimated the distance to all galaxies and clusters by a factor of ~7.8. (Reproduced by permission of the A.A.S.: E. Hubble & M. Humason (1931). Astrophys. J., 74, pp. 43–79.)

changes in the galaxy distribution across several adjacent photographs. One of the new all-sky surveys came from Lick Observatory and is referred to as the Shane and Wirtanen galaxy study. The second came from Mt. Palomar Observatory with funding from the National Geographic Society. Two prominent analyses of galaxy clusters were completed from the National Geographic Palomar Observatory Sky Survey, one from Caltech professor Fritz Zwicky and the second from UCLA astronomer George Abell. Both of these cluster catalogues contain thousands of rich galaxy clusters, each somewhat resembling (but outshining) the more local Virgo cluster of galaxies that sits near the center of our Local Supercluster. Abell went further than Zwicky to assert that his rich galaxy clusters gather together in pairs or triplets to form superclusters, even though this claim remained controversial until galaxy redshift surveys reached full stride. The main point here being that by 1975 the concept of superclusters had been extant for more than two decades, and included in these studies were detailed maps by de Vaucouleurs of the Local Supercluster. On the other hand, there were astronomers and cosmologists who doubted the existence of superclusters – including the reality of our own Local Supercluster – despite the evidence presented by Shapley, Shane and Wirtanen, de Vaucouleurs, and Abell. The ground was fertile for new discoveries.

2

Preview of the Discovery of Cosmic Voids

The discovery process documented in this book is multifaceted and deserves a detailed step-by-step explanation. Advances were being made simultaneously by both observational astronomers and theoretical astrophysicists as the process moved forward. There were skeptics who aimed to hold tight to their traditional concepts, and there were open-minded explorers like me and my close collabora- tor Steve Gregory who began to see an entirely new view. This chapter aims to present the breakthrough advances without getting bogged down in details. The details of the discovery absolutely do matter, but the initial story needs to be told first, so the details that are explained in later chapters can be better understood and placed in an appropriate context. This chapter is therefore a first look at the initial discovery as well as a guide to later chapters. Some might see this chapter as an extended preface that previews the entire series of published papers – both theoretical and observational – as laid out in the timeline of discovery presented in Table 8.2. Those who wish to do so might want to flip from any narrative in the various chapters to this timeline and use it as a touchstone. The process of laying out the facts should help resolve persistent priority disputes that have existed in this field of study for 35–40 years.

2.1   Revealing a Hidden Paradigm

In 1975, there remained an undiscovered and hidden paradigm: in addition to the known galaxy superclusters, there were huge empty regions of space that cumulatively occupy 60–70% of the volume of the Universe. These cosmic voids sit between and around the filamentary supercluster structures that contain the galaxies. Today, it is abundantly clear: galaxies are not homo- geneously distributed on the scale of the void and supercluster structure.

However, no astronomer had thought of searching for vacant 3D regions in the galaxy distribution. What appeared to the earlier astronomers like Holmberg, Hubble, and Zwicky as a uniform "field" of galaxies was actually a superposition along the line of sight of sheets and filaments of galaxies that connect one rich cluster to another. In traditional 2D galaxy maps, the cosmic voids, the filaments, and the sheets of galaxies were indistinct, and it would take a full 3D analysis of the galaxy distribution to delineate the complex structure.

The edifice – built by Hubble and Zwicky some 40 years earlier – describing a homogeneous "field" of galaxies became vulnerable in the late 1970s, because a new technology had made its way to large telescopes. Astronomers' reliance on photographic plates was shifting to hybrid systems that added a twentieth-century-era electronic image intensifier into the camera system. The intensifiers were 10 times more efficient at capturing the light of astronomical objects than plain photographic emulsions. I shall describe, in Chapter 5, that this advancement made a huge difference to astronomers like me and Stephen A. Gregory (b. 1948). We aimed to collect and measure as many galaxy spectra as possible to make the first wide-angle 3D maps of the galaxy distribution surrounding and connecting rich galaxy clusters. With these new spectrograph systems, a small group of astronomers began to collect a flood of new data and along with it provided distance estimates for many hundreds of additional galaxies. It was in our very first 3D map that enormous cosmic voids in the galaxy distribution were first clearly revealed, as were suggestions that bridges of galaxies connect large galaxy groups and clusters. The bridges we detected would later be identified with filaments and/or sheets of galaxies in the cosmic web.

The new electronic detectors were developed for astronomy over a 10-year period starting in the mid-1950s, with pioneering work done at the Observatoire de Paris in France (by André Lallemand), at Imperial College of the University of London (by J. D. McGee), and at Lick Observatory in California (by G. Kron). The first systems incorporated evacuated glass chambers that had to be broken open to retrieve energy-sensitive emulsions used to record accelerated electrons. Soon, a new design emerged, incorporating a separate electronic image intensifier tube (with a green-colored phosphorescent amplified output image) plus an ordinary photographic plate to record the amplified image. With this design change, the intensifier systems became practical for general use in astronomy. The pioneering group that introduced the newer systems into observatories was located at the Carnegie Institute of Washington, in a division called the "Department of Terrestrial Magnetism" (DTM). The scientist-in-charge was W. Kent Ford, Jr. (b. 1931). His work came to fruition in the mid-1960s, and during the later development phase, he worked on astronomy research projects with another DTM scientist, Vera Rubin (1928–2016). Together, Rubin and Ford (1970) first measured the rotational motion in the far outer parts of the Andromeda galaxy, and then, by 1978, extended their study of galaxy rotation to include many spiral galaxies. Based on these results, they are credited with showing the dominance of dark matter in the outer halos of many ordinary galaxies.[1] Image-intensified detectors were crucial for their success.

By the early 1970s, spectrographs with image-intensified detectors were already operating on several telescopes in Arizona and in Texas. DTM assembled a "visiting" system that was used at Lowell Observatory near Flagstaff and at Kitt Peak National Observatory (KPNO) near Tucson. KPNO acquired a system of its own, as did the University of Arizona's Steward Observatory (all for telescopes located on Kitt Peak). Gregory and I entered graduate school at Steward Observatory in the 1969/1970 academic year, at a time when a new Steward Observatory spectrograph system was just coming into operation. We both selected as our PhD thesis advisor William G. Tifft, the Arizona faculty member who worked most closely with the new spectrograph. Gregory was one of two graduate students first permitted to use the image-intensified spectrograph at the Steward Observatory 90-inch telescope (the other was our fellow student, Leo Connolly). In that era, the goals were simple: record the spectrum of a galaxy and measure the faint features in the spectrum to obtain the galaxy's Doppler velocity, thus revealing an inferred distance; and do that, one galaxy at a time, for as many nights as possible. An efficient astronomer could record up to 20 galaxy spectra per night. The combination of successful observing programs both at the KPNO 84-inch telescope and at the Steward Observatory 90-inch telescope on Kitt Peak (Figure 2.1) began the flood of new galaxy redshifts that would put within our grasp the first deeply probing 3D maps of the galaxy distribution. These redshift survey programs at Kitt Peak were started seven years before the closest competitor (from 21-cm radio wavelength measurements at Arecibo Observatory) and eight years before the commissioning of the so-called redshift machine built at Harvard's Center for Astrophysics.

Soon after Gregory and I finished our PhD degrees at the University of Arizona, we began a joint effort and worked with one of the intensified spectrograph systems at the KPNO 84-inch telescope. This required us to be in top-rank competition with other astronomers who aimed to do their own work with the KPNO telescopes (including Rubin and Ford). Two other scientists, Guido Chincarini and Herbert Rood, were collecting spectra of galaxies from KPNO in this same era, but relative to their allotted telescope time prior to 1974, they had published a limited number of papers. In 1975, Steve Gregory and I submitted a proposal for telescope time at KPNO and adopted a novel strategy

Figure 2.1 Kitt Peak, Arizona. Three telescopes in this photograph are part of the story discussed in this book. The large white dome, third from the left, houses the KPNO 84-inch telescope, the largest dome in the picture (on the far upper right) houses the KPNO 150-inch telescope, and the tall cylindrical structure immediately to the lower left of the 150-inch telescope houses the Steward Observatory 90-inch telescope. These are now, respectively, called the 2.1-m, the 4-m Mayall Telescope, and the 2.3-m Bok Telescope. Kitt Peak is located within the Tohono O'odham Reservation, and the telescopes sit on land leased from the reservation by the National Science Foundation (NSF). By permission: copyright NOAO/AURA/NSF.

(our original 1970s-era observing proposal is shown in Appendix A). We aimed to complete a galaxy redshift survey that spanned a wide swath of sky 24° long stretching between two rich galaxy clusters, Coma and A1367, in order to view the 3D galaxy distribution over a large angular scale. This had not been done before. Our proposal followed the scientific method, in the sense that we asked very specific questions. For example: Are these two rich clusters connected by a bridge of galaxies? Are they located within a common supercluster? Our proposal was successful; we collected and analyzed 44 new galaxy spectra yielding a total survey sample of 238 galaxies in and around these two rich galaxy clusters, and we mapped the 3D galaxy distribution over the 24° wide area, thereby producing the largest continuous angular survey at that time. It was our Coma/A1367 redshift survey (Gregory & Thompson 1978) that provided the most convincing early demonstration (see Sandage 1987) that the local galaxy distribution is highly inhomogeneous in a 3D view, including galaxy enhancements (small groups and clusters of galaxies) as well as significant deficiencies (cosmic voids). This ended Hubble's historic hammerlock on the accepted view of the galaxy distribution.

The biggest surprise was the discovery of cosmic voids, the enormous empty regions that sit in the foreground of the Coma/A1367 supercluster. We were the first to use a 3D redshift map to identify and measure the diameter of these vast empty regions; we were also the first to use the word "void" in this context in the astronomy literature. Cosmic voids have diameters greater than 90 million light-years, or as astronomers write it, >20 h^{-1} Mpc. Appendix B contains a reproduction of the final published version of our 1978 scientific paper entitled "The Coma/A1367 Supercluster and Its Environs," which reports the discovery of cosmic voids.

In our 3D wide-angle redshift map, Gregory and I also identified a filament of the cosmic web (the first extended contiguous structure located far outside the Local Supercluster) stretching between the two clusters, Coma and A1367. We called it a "bridge" between the two rich clusters. In follow-up work, Gregory and I immediately began to search for similar bridges or filaments between other rich Abell cluster cores and eventually confirmed that all of the richest nearby galaxy clusters are embedded in distributed superclusters. Outside of the extended supercluster systems, we repeatedly confirmed the presence of huge cosmic voids. The main redshift map from our 1978 paper is shown in Figure 2.2. Further details of both the discovery and our extensive follow-up work are presented in Chapter 5 and summarized at the beginning of Chapter 8.

The Gregory and Thompson Coma/A1367 supercluster manuscript was received by the *Astrophysical Journal* on September 7, 1977. This date is significant because a group of astronomers from Tartu Observatory in Estonia, at that time part of the USSR, was simultaneously studying the distribution of galaxies and clusters of galaxies, searching for clues that might reveal evidence related to galaxy formation. They had organized an international scientific meeting on this topic, scheduled for September 12–16, 1977 (a meeting that neither Gregory nor I attended). As described in Chapter 6, the Tartu Observatory astronomers played a subsidiary role in the discovery of cosmic voids and the network of filamentary structure in the galaxy distribution. The Gregory and Thompson paper was received by the *Astrophysical Journal* essentially in its final form before their conference began and five months before the Estonian group had a chance to publish their own results. Even though they were on the right track, they had insufficient data to prove their case for what they called "holes" in the galaxy distribution. In an era when cosmologists and many astronomers were ready to reject the introduction of any added complexity into the discussion of the galaxy distribution, the Tartu Observatory group was in need of additional observational proof to strengthen their case. It was our "complete" galaxy redshift surveys that proved the central concept. Although their ideas were circulated as preprints at the September 12–16

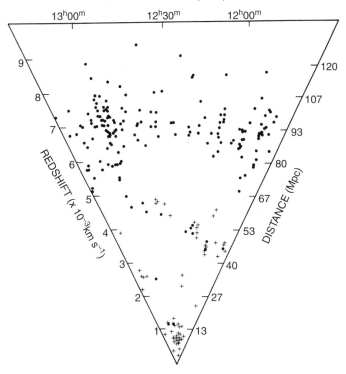

RIGHT ASCENSION (1950)

Figure 2.2 Galaxy redshift survey of the Coma/A1367 supercluster. Each small, black symbol in this plot represents a single galaxy with a measured redshift. After collecting our galaxy redshift measurements, we applied the Hubble-Lemaitre velocity–distance relation to obtain distance estimates for each galaxy and plotted them as a function of the angular position on the sky (right ascension). The Milky Way galaxy – including Earth – sits at the bottom apex of the diagram with the Universe stretched out above us. Ideal cone diagrams are three-dimensional, but for the sake of simplicity, the cone is viewed from one side, so our 24° × 15° survey volume is displayed as a wedge spanning 24°. This diagram was the first to display (for a well-defined complete sample) a wide-angle 3D view of the galaxy distribution of the deep Universe showing definitive evidence for huge cosmic voids and suggestions of filamentary structure. The elongated dense clump of points on the left (at 13^h 00^m and a distance of 95 Mpc) is the Coma cluster of galaxies. A less dense clump of points on the far right (again at a distance of 95 Mpc) is the galaxy cluster A1367.

A horizontal "bridge" of scattered points connects the two clusters. Two cosmic voids dominate the foreground volume, one on the near-side of the supercluster "bridge" and the second displaced to the left of center, at a distance of ~45 Mpc. Keep in mind that the detected galaxies contribute only a small fraction of the density of the Universe, but they light up the regions where CDM has accumulated in what appears (even in this pioneering diagram) to be filament-like features. Hubble and Zwicky would have expected this volume to have been uniformly populated with galaxies, except for the elongated clumps at the locations of Coma and A1367. By permission of the A.A.S.: S. Gregory & L. Thompson (1978). *Astrophys. J.*, 222, 784–99.

scientific meeting, the Tartu Observatory astronomers also submitted their analysis to the astronomy journal *Monthly Notices of the Royal Astronomical Society*. The journal gave their paper careful scrutiny, and after a lengthy period of challenges by the referee and major revisions by the authors (including adding the best references they had at the time acknowledging our work from galaxy redshift surveys), their paper was eventually published (Joeveer, Einasto, & Tago 1978). In this regard, the redshift surveys from the Arizona telescopes remained the central early proof of the cosmic void phenomenon and the demonstration that bridges of galaxies connect rich galaxy clusters. The interplay between the results of the Tartu Observatory group and our Coma/A1367 study is discussed in more detail in Chapter 6.

As noted earlier, the same spectrograph technology used to discover cosmic voids was also used by Rubin and Ford to reveal dark matter halos that surround all spiral galaxies, and our paper on cosmic voids and their primary paper on dark matter (with N. Thonard) were both published in 1978. This is a fitting coincidence because these two concepts are intimately linked through the LCDM model: the dark matter that dominates galaxy halos collapses earlier in time (as well as on larger scales) than the ordinary matter in individual galaxies. The dark matter forms the backbone of the supercluster structures, and as the supercluster structures defined by the CDM become enhanced by gravitational infall, the underdense regions are drained, leaving behind empty voids. The Rubin and Ford and the Gregory and Thompson papers were the opening shots of a revolution that has consumed the attention of astronomers and cosmologists for the past 40 years, as they build models to simulate how galaxies and clusters of galaxies form and evolve.

Gregory and I were fortunate to participate in these discoveries. Our work followed shortly after we completed our formal graduate school training and benefitted from indirect mentoring by many key scientists in this era. Quietly and often in the background of these discoveries was UCLA Professor George Abell, a venerable pioneer in the study of galaxy clusters, superclusters, and the large-scale structure of the Universe. A graduate class I took in galaxies and cosmology from George Abell in the spring of 1969, when I was still an undergraduate student at UCLA, launched my early career in astronomy. Gerard de Vaucouleurs and his wife Antoinette were guest lecturers in a graduate seminar class I took at the University of Arizona in the early 1970s. Most impressive and influential to me was another great scientist from this era, Jan Oort, whom I met and with whom I talked a number of times at scientific meetings. I admired his ability to identify kernels of truth that are often found buried in a cluttered field of ideas. Oort summarized the status of super-clusters in a cogent and carefully written review article published in 1983. His review provides a crisp snapshot of the very rich and developing concepts at a time when the early studies of the large-scale structure were still in their formative stage. Allan Sandage, Edwin Hubble's protégé at the Carnegie Observatories in Pasadena, was a mentor and personal supporter of the breakthrough that Gregory and I uncovered. He and I spoke any number of times about this and about other scientific issues. Finally, the impressive theorist Yakov Zeldovich, whom I met and spoke with just once – in late August 1982 at the International Astronomical Union (IAU) Meeting in Patras, Greece – was, in my view, a brilliant theoretical physicist whose role as a provocateur is described in this book. Unfortunately, all five of these great scientists are no longer among the living.

Stephen A. Gregory (b. 1948)

Steve Gregory was born and raised in central Illinois and attended the University of Illinois at Urbana-Champaign as an undergraduate student, at a time when the Astronomy Department was housed in the Old Observatory Building. At that time, the head of the Astronomy Department was the eminent British mathematician and cosmologist Professor George McVitte. In early 1970, Gregory began graduate studies at the University of Arizona where he worked first with Ray Weymann, and later on his PhD degree with William G. Tifft measuring hundreds of galaxy redshifts in the Coma cluster of galaxies. In 1973, Gregory received his first faculty appointment at the State University of New York, Oswego. This position was the only openly advertised professor appointment available that year in the United States, as the country was in the midst of a major recession. Gregory later moved to Bowling Green State University in Ohio, and from there to the Physics and Astronomy Department at the University of New Mexico in 1984. In 2007, he retired from the UNM Astronomy Department to become a Senior Scientist for Boeing (which supports the US Air Force Research Laboratory), specializing in the photometric characterization of Earth-orbiting satellites for defense purposes. Gregory currently lives in Albuquerque, New Mexico. Photo reproduced by permission: copyright Stephen Gregory.

Over the years, other participants in this story have shared their experiences and have written about the roles they played in the discovery of the large-scale distribution of galaxies. In this book, I integrate many of these contributions with my own views and experiences to produce a comprehensive documented description of the discovery process. Two of my former collaborators each have contributed their stories: the late Herbert J. Rood and his close collaborator Guido Chincarini (Rood 1988a, 1988b; Chincarini 2013). The late John Huchra wrote his own brief biographical summary, and many of his experiences are also discussed in an interview recorded by the American Institute of Physics. In these sources, he described his work on galaxy redshift surveys (Huchra 2002). Huchra's primary collaborator Margaret Geller has written many shorter pieces that present her perspective including one I discuss at length in Chapter 8. J. Richard Gott III presents in his book entitled *The Cosmic Web*, published in 2016, the efforts he and his collaborators made to understand the topology of the large-scale structure, less than a decade after the earliest discoveries. Simon White, one of the pioneers in the construction of N-body computer simulations that show how galaxies and the large-scale structure in the galaxy distribution grow and evolve over cosmic time, summarized his contributions in his acceptance address for the Shaw Prize (White 2017). In 2013, Jaan Einasto of Tartu Observatory in Estonia wrote a lengthy monograph entitled *Cosmic Web and Dark Matter Story* that describes his work, both on the dark matter halos that surround galaxies and on the effort initiated by Yakov Zeldovich that led to a near-simultaneous detection – relative to the Gregory–Thompson work – of cosmic voids by scientists at Tartu Observatory. Einasto wrote another shorter summary of his own work five years later (Einasto 2013, 2018).

2.2   Theoretical Models of Galaxy Formation: A Brief Introduction

In 1970, Zeldovich (1914–87) published an interesting and widely acknowledged paper describing how material in the Universe slowly gathers together (astronomers call this a "gravitational collapse") as the matter responds to gravity (Zeldovich 1970). He and his students applied these concepts to the raw material from which galaxies begin to form in the early Universe, after it has sufficiently cooled. Their first version of these models was naive because dark matter played no role, so only ordinary baryonic matter was considered. The collapsing matter in these Zeldovich models formed huge extended gaseous superclusters that Zeldovich and his successful student Rashid Sunyaev (b. 1943) suggested would subsequently fragment into galaxies. Simultaneously, the Zeldovich models predicted that large vacant holes would develop in the matter distribution. Jaan Einasto (b. 1929) tells the story in his monograph that, as Zeldovich made progress on this theoretical work, he enquired at Tartu Observatory, hoping to learn about the observed spatial distribution of galaxies and galaxy clusters. The astronomers at Tartu Observatory did not quite know what to do at first, but without access to new observations, they were left to study existing catalogued samples of galaxies and clusters of galaxies. They ended up with sparse maps of the galaxy distribution because the redshift samples available in existing published catalogues in 1975–77 were few and far between. They began to see some evidence for large-scale irregularities in the galaxy distribution, but without complete red-shift survey data, even when they saw an apparently empty volume, they could not prove it was a true cosmic void (as opposed to a region that had not yet been included in the sparsely sampled redshift studies available in the published literature).

A description of the work at Tartu Observatory in Chapter 6 provides a natural transition to Chapter 7, where I present a lengthy discussion of the theoretical developments that were occurring in the period 1970–90. At this time, the hierarchical model of galaxy formation championed at Princeton University by P. J. E. Peebles (b. 1935) was being challenged by Yakov Zeldovich and his students at the Institute of Applied Mathematics in Moscow. The Zeldovich "pancake" model of supercluster formation was continuously modified during this period in order to fit constraints placed on it by new results from astronomical observations and particle physics, and new developments were reported for the hierarchical model as well. Although the "pancake" model at first seemed best suited to explain the formation of cosmic voids and the network of sheets and filaments in the galaxy distribution (even Oort tilted in this direction in his 1983 review), eventually the original form of the gaseous "pancake" model stumbled, after which CDM entered the scene. Finally, what is now called the "Zeldovich approximation"[2] was applied as the starting condition of all the newer galaxy formation models in that era, and the figurative pendulum swung back to favor the hierarchical model. This is an ironic outcome because, in the preceding years, Peebles worked to question the reality of observational evidence for filamentary structure that appeared in the galaxy distribution. He suggested, repeatedly, that the eye has a tendency to see false structure in what he believed were random dis-tributions of objects. Peebles favored interpreting observational phenom-ena in terms of uniformity and homogeneity that supported hierarchical concepts. On the other hand, those who were developing the early Zeldovich model of supercluster formation were more open-minded and quicker to accept the large-scale structure observations that showed fila-ments and sheets in the galaxy distribution.

Based on the order of structure formation, the original Zeldovich models were called "top-down." This stood in direct contrast to the hierarchical galaxy formation models in the West that assumed "bottom-up" galaxy formation: the smallest structures collapse first to form stars and star clusters, and then these smaller objects gather together, piece by piece, to build large galaxies and eventually clusters of galaxies. Zeldovich and his collaborators pursued the top-down model for seven to eight years, before CDM entered the scene and changed everything. While this was happening in the early 1980s, Adrian Melott (b. 1947), a PhD graduate from the University of Texas and a young expert in the physics of neutrinos and models of galaxy formation, visited both Tartu Observatory in Estonia and the Zeldovich research group in Moscow, where Sergei Shandarin (b. 1947) and Anatoly Klypin (b. 1953) were also doing pioneering work, building computer models to simulate the process of supercluster collapse. Soon the original gaseous form of the "Zeldovich pancake model" was left to history to be replaced by newer models incorporating neutrino dark matter. At first, the extended dark matter neutrino structures were identified as "Zeldovich pancakes," but they were really quite different from the early Sunyaev and Zeldovich concept, in the sense that they did not provoke a catastrophic gaseous collapse and immediate fragmentation into smaller galaxy-sized pieces. These newly conceived dark matter structures did have the advantage, however, of being able to begin forming at an early epoch, somewhat before any gaseous objects could collapse on their own.

In 1983, soon after Melott and his Russian collaborators tested neutrino dark matter in N-body computer simulations, they pushed the frontier further forward and tested CDM as an alternative to neutrino dark matter. (Neutrinos travel at velocities near the speed of light whereas CDM particles move much slower. Accordingly, the former is called "hot" and the latter "cold.") The CDM results showed a much closer match to the observed cosmic void and supercluster distribution based on the early 3D redshift survey maps produced by telescopes in Arizona. This was the first indication that the top-down models of galaxy formation, which originally seemed to show the formation of Zeldovich pancakes, could somewhat seamlessly be transformed into a mechanism for forming a superstructure of dark matter halos that allowed the ordinary matter to accumulate in the halos from the bottom-up, in a hierarchical fashion. At about this same time, a group of Western scientists entered the scene and built more refined N-body computer models that eventually eclipsed the pioneering work done in Moscow and the work spearheaded by Adrian Melott in the United States. The new group included Marc Davis (b. 1947), George Efstathiou (b. 1955), Carlos Frenk (b. 1951), and Simon White (b. 1951). Chapter 7 provides a detailed and more complete account of the historical work on models that describe the formation of structure in the galaxy distribution, where the work of this group is often abbreviated as DEFW.

A central issue in adopting any major transformational concept is whether observations of a particular phenomenon can be understood in terms of a reasonable theoretical model. The famous British astrophysicist Sir Arthur Eddington (1882–1944) summarized it as follows:

> Observation and theory get on best when they are mixed together, both helping one another in the pursuit of truth. It is a good rule not to put overmuch confidence in a theory until it has been confirmed by observation. I hope I shall not shock the experimental physicists too much if I add that it is also a good rule not to put overmuch confidence in the observational results that are put forward until they have been confirmed by theory. (Eddington 1934)

The enormous diameters of the cosmic voids Gregory and I first reported in 1978 were far beyond what many theoretical cosmologists in the western hemisphere considered plausible inhomogeneities, and some of these cosmologists seemed perfectly satisfied to reject our observations, until the concept of CDM was integrated into models of galaxy formation in the mid-1980s and our observations were reconfirmed with both optical and radio telescope observations. The process of building an acceptable theoretical explanation for cosmic voids culminated in 1996, when a paper written by J.R. Bond, Lev Kofman, and Dmitry Pogosyan (Bond et al. 1996) presented the "peak-patch" model. It incorporates early hierarchical dwarf galaxy and star formation on small scales and combines it with the growth of dark matter-dominated cosmic structure on the largest scales. Just like the pioneering work of Shandarin, Klypin, and Melott, their initial conditions for the dark matter dynamics are based on the Zeldovich approximation. The model by Bond and his collaborators also utilized a concept called "biased galaxy formation" introduced by Kaiser (1984). The bias and peak-patch concepts mean that galaxies (and their stars) form preferentially in the highest peaks of the 3D distribution of dark matter density irregularities, and these high density peaks themselves preferentially congregate into contiguous patches, with a relative contrast (with respect to the general background distribution of galaxies) that grows with time. This is why the term "bias" is used. In the peak-patch model, the era of supercluster formation is now. In other words, if we use telescopes to look back into the distribution of galaxies at earlier times, the relative contrast of the supercluster structure will be somewhat diminished, compared to what we see today. The irregularities that lead to the dark matter web-like structure may well have their origin in quantum fluctuations in the very early Universe. If so, these fluctuations were stretched and expanded during an inflationary era from microscopic dimensions to huge scales, some of which are so large that they extend beyond our current horizon. Details like these are now being discussed in ongoing research.

2.3   The Discovery Process

After major scientific discoveries are made, the new concepts need to be assimilated into the body of scientific knowledge. This proceeds by an ill-defined process that merges the new ideas with earlier ones, thus creating a new consensus view. Steven J. Dick (2013) described in his book entitled *Discovery and Classification in Astronomy* many examples demonstrating that discovery is a process. Indeed, the discovery of the cosmic web was an extended process, and this is made abundantly clear in the lengthy timeline of discovery presented in Chapter 8. The early reports of cosmic voids published in 1978 triggered a discovery process in the same manner that Penzias and Wilson's early observations of a uniform microwave signal across the sky in Holmdel, New Jersey, triggered the discovery process of the complexities associated with the cosmic background radiation.

In 1978, Gregory and I – and the others around us – immediately realized the significance of our early galaxy redshift survey maps. We worked as quickly as we could to confirm the cosmic void and supercluster phenomenon in and around other groups of galaxy clusters (i.e., superclusters), by publishing additional redshift surveys in both the northern and southern galactic hemispheres. In early 1982, we described our new results in the semi-popular magazine *Scientific American*. Simultaneously, the efforts started by Zeldovich and his collaborators and by Peebles pushed forward in a theoretical vein. This entire body of work – along with the dark matter studies of Rubin and Ford – would eventually merge into a coherent picture to define the cosmic web modeled by N-body computer simulations, a discovery process that was essentially complete by the early 1990s.

Somewhat before major discoveries are made, it is not uncommon to find precursor studies that hint at or come close to the new result. One or even a series of early revelations can occur. At first, these early results may appear disconnected. Once the break-through discovery is announced and the new paradigm is clearly stated, the early disconnected ideas are clarified. This happened when the cosmic microwave background was identified in 1965, and it happened again in the late 1970s when cosmic voids were discovered. Chapter 8 documents the chronology of relevant events, including the precursors.

Laird A. Thompson (b.1947)

Laird A. Thompson was born in Lincoln, Nebraska, but moved with his family to Southern California for high school and college. After graduating cum laude from UCLA with majors in both Physics and Astronomy, he attended the University of Arizona for graduate studies in observational astronomy and was awarded a PhD in November 1974. After a two-year research appointment at Kitt Peak National Observatory, he started as an entry-level professor at University of Nebraska, Lincoln, where he worked for 2 1/2 years, before moving to University of Hawaii's Institute for Astronomy for eight years. During his first 18 months in Hawaii, he laid a complete foundation for a distant supernova search with the newly commissioned 3.6-meter Canada-France-Hawaii Telescope. (This coincided in time with a separate effort by S. Perlmutter of Lawrence Berkeley National Lab, who was starting his own distant supernova search.) When the National Science Foundation (NSF) did not support Thompson's distant supernova search, he switched fields to do pioneering work in laser-guided adaptive optics including the first experimental projection of a 589 nm sodium wavelength laser from Mauna Kea (launched from the current site of the Gemini North Telescope and published in *Nature* as Thompson and Gardner 1987). His laser experiment collaborator, Illinois Electrical and Computer Engineering Professor Chester Gardner, was key to Thompson joining the Astronomy faculty at University of Illinois Urbana-Champaign. With funding primarily from NSF Advanced Technologies and Instrumentation, Thompson led an adaptive optics group from Illinois to build and operate a laser-guided adaptive optics system at the coudé focus of the Mt. Wilson 100-inch telescope, the same telescope Shapley and Hubble used early in the 20th century for pioneering work in cosmology. In 2014, Thompson switched to an Emeritus Professor appointment. Photo reproduced by permission: copyright Laird A. Thompson.

The combined timeline in Chapter 8 is especially useful for seeing the Gregory and Thompson work in relation to other redshift surveys that were published in the years that followed ours. The two most important subsequent surveys came from Harvard's Center for Astrophysics (CfA). When the CfA1 survey results were published in 1982, the authors referenced our redshift work up front, acknowledged our contribution, and called their new survey "complementary" to the existing narrower but deeper redshift surveys we had published (Davis et al. 1982). Gregory and I prefer an alternate description: the CfA1 survey undersampled the 3D galaxy distribution and, within the limits of their data, confirmed our original discovery. But as the years have passed, the story has morphed and quite often references to our work have been omitted entirely. I cite as an example a recently published historical discussion by Marc Davis, who was head of the CfA1 survey in the early 1980s. In this recent presentation, Davis (2014) calls his own work the "first redshift survey of galaxies," describes the 3D structure with voids and superclusters, but makes no reference whatsoever to the Gregory and Thompson discovery work.

Next in importance is the 1986 CfA2 study by Valerie de Lapparent (b. 1961), Margaret Geller (b. 1947), and the late John Huchra (1948–2010) that also had a significant overlap with our 1978 cosmic voids paper and that, sometimes, is mistakenly credited with showing the first observational evidence for cosmic voids and the detection of structure in the galaxy distribution (de Lapparent et al. 1986). When this happened, Geller and the late Huchra did little to nothing to correct the misconceptions. The timeline in Chapter 8 shows the development sequence of the early redshift surveys. The role of the de Lapparent et al. study was to extend our 1978 results (and those of the CfA1 survey) by presenting a wider-angle and somewhat deeper survey that included fainter galaxies. This sharpened the visibility of the structure we had already identified. De Lapparent et al. (1986) also provided evidence for supercluster structure on a scale that was as large as their survey volume. But as the timeline in Table 8.2 documents, the CfA2 studies were being done at the close of the pioneering period of redshift surveys. It is also significant that CDM models of galaxy formation had already been introduced and were under active development before the first CfA2 papers were published.

Once these pioneering studies demonstrated the great rewards that came from extending redshift surveys into the deeper Universe, redshift survey programs began to proliferate. The key contributors at this point became the Las Companas Redshift Survey, the Two Degree Field (2dF) Galaxy Redshift Survey, and the Sloan Digital Sky Survey. These are all described in later chapters and are included in the timeline shown in Table 8.2. For those who want to see a preview of the rich rewards, look ahead to Figure 5.8, where I show graphically the progression from the Gregory and Thompson (1978) map to the CfA2 Slice of the Universe map and finally the Sloan Digital Sky Survey results reprinted from the publication *A Map of the Universe* by Gott et al. (2005).

This book closes with a discussion in Chapter 9 of specific contributions made to both astronomy and cosmology that were triggered by the identification of cosmic voids. Standing high above all other accomplishments – and reinforcing the central theme of this book – is the recognition that cosmic voids in the local galaxy distribution provide a touchstone or tool for those who build N-body simulations to test and refine the dark matter models as they relate to galaxy formation and the large-scale structure. Simon White recently accepted the 2017 Shaw Prize for his contributions to the general understanding of structure formation in the Universe. In his acceptance address, he described how the appearance of cosmic voids provided the means to decide that CDM was favored over neutrino dark matter. White (2017) went on to state: "The demonstration that no known particle [from the Standard Model of particle physics] can account for the dark matter remains one of the most significant contributions of computer simulations to astrophysics and cosmology." The foundation for White's work and for his recent award was the discovery of cosmic voids as described in this book.

2.4   Working with Cosmic Voids

In the past 35–40 years, attempts have been made to construct catalogues and to analyze the statistical properties of cosmic voids. These studies reveal basic void characteristics: diameters, 3D shapes, and average enclosed underdensities. Defining voids is not a simple process because occasionally one or even several isolated galaxies reside inside an otherwise enormous empty volume, and because the topology of the void-supercluster structure is sponge-like, one void can merge into the next. So rather than identifying completely empty volumes, it is appropriate to set an upper limit on the enclosed galaxy density (number of galaxies per cubic megaparsec or the number galaxies per cubic light-year) and to consider identification methods that involve hierarchies and connected volumes. As described in detail in Chapter 9, various techniques – from the simplest to the most elegant – have been used to catalogue cosmic voids. During this process, lists of isolated "void galaxies" have also been compiled. Astronomers have already studied large numbers of void galaxies and, based on these studies, have addressed questions about the broader issues of galaxy formation and galaxy evolution.

Another line of enquiry involves the topology of the large-scale structure and the nature of the underlying perturbations that led to the structure. A fair number of scientists have pursued topological studies, one of the more prominent being J. Richard Gott III. His 2016 popular-level book entitled *The Cosmic Web* describes his own work in this area of study. At times, he simplifies the question of topology by giving examples such as an empty space filled with "meat balls" where galaxies are concentrated in isolated structures. Another example is a space filled with galaxies that incorporates embedded voids somewhat like the holes in Swiss cheese. A third example is the topology similar to that of a sponge, where empty voids are surrounded by walls and filaments populated with galaxies. As Gott's book (and his published work) makes clear, both the topology of the large-scale galaxy distribution and the nature of the original underlying perturbations may provide information relevant to the early inflationary epoch of the Universe.

Cosmologists have come to realize in the past decade that cosmic voids provide an ideal setting to probe the nature of dark energy, the component of the Universe that represents 69% of the total energy content of the Universe. The reason voids are useful in this regard is easy to understand. The physical processes that occur in the denser regions of the Universe – those surrounding galaxies and galaxy clusters – are complex because the baryons (the ordinary matter in stars and galaxies) interact with other baryons as they fall into the dark matter halo of a galaxy or into the halo of a rich cluster. It is difficult to disentangle the complex interactions of the baryons from the generally more subtle effects of dark energy. Cosmic voids, on the other hand, have a simpler history, and the way cosmic voids respond to the dark energy turns out to be more straightforward. As huge galaxy catalogues come available from new galaxy redshift surveys, some of which are not yet completed, the dynamics of expanding voids may tell us whether dark energy is a plain and simple mani-festation of Einstein's cosmological constant or whether it is a more compli-cated phenomenon. All matters that pertain to how cosmic voids inform modern cosmology are covered in Chapter 9.

2.5   Moving On to the Full Story

The fully documented discovery story begins in Chapters 3 and 4 by first addressing the historical question as to how astronomers managed to overlook for so long the fact that 60–70% of the volume of the Universe contains essen-tially no galaxies whatsoever. What historical events set up this conundrum? Already mentioned is the fact that studies of the 2D distribution of galaxies on the sky obscure most evidence of the remarkable 3D irregularities. Then there is a tendency – even a principle – in science to favor the simplest model or the simplest paradigm. When Hubble initially explored the galaxy population in the immediate environs of our Milky Way, he found it to be generally uniform because we happen to live in the outskirts of the Virgo supercluster – often called the Local Supercluster – a huge flattened structure that is, more or less, uniformly populated with average galaxies. If we look far enough away from the plane of this flattened structure, indeed we can detect striking evidence of inhomogeneities. In the 1930s, Harlow Shapley did exactly this, but Hubble did not trust Shapley's results and spent no time checking for irregularities on these intermediate scales. Instead, he probed along numerous "pencil beams" that stretched deep into the distant Universe and simply ignored all irregularities along the line of sight. Once the general concept of homogeneity was set in the minds of astronomers and cosmologists in the 1930s, it was difficult to dislodge even though evidence to the contrary would appear now and then. The evidence would be acknowledged, but no new consensus view would take root. For example, Chapter 4 discusses how numerous galaxy superclusters were discovered by Shane and Abell in the 1950s and 1960s, and how our own Local Supercluster was studied by de Vaucouleurs starting in 1953, but because of Hubble's early influence, those who analyzed the galaxy distribution prior to 1975 assumed most often that superclusters were embedded in a uniform field of galaxies. When evidence was presented in the 1950s for a model where all galaxies are situated in clusters, the study was acknowledged but then ignored. It was Gregory's and my privilege to break the old paradigm with our first complete wide-angle redshift surveys and to point the way to the modern view with 3D maps that show cosmic voids dominating the overall structure of the Universe. But we were among only a small group of astronomers who pushed this idea to the fore. The person some have called "one of the fathers of modern cosmology," Professor P. J. E. Peebles at Princeton University, ignored our work for a decade and suggested that those who saw filaments in the galaxy distribution were misleading themselves. These tensions and the manner in which they were resolved make for an interesting story.

3

Homogeneity of the Universe: Great Minds Speak Out

Among the most fundamental questions in cosmology is whether matter is distributed homogeneously over large scales in the Universe. Its corollary is isotropy: Does the Universe look the same in every direction? For example, if significantly more mass is seen in the distant Universe in one hemisphere than in the opposite hemisphere, the Universe is anisotropic as well as being inhomogeneous at the scale used in the test. Starting as early as 1917, Einstein assumed both homogeneity and isotropy when he built his early static model of the Universe, and other cosmologists have often done the same. This assumption is deeply rooted in cosmology. It was given the formal name "The Cosmological Principle" by Milne (1935). Despite his use of the word "principle," it is widely acknowledged to be an assumption that is subject to observational test.

On small scales like those of our Solar System and stars in the Milky Way, the distribution of mass is decisively clumped and is, therefore, inhomogeneous. On the very largest scales observed – like the distance to our horizon at 13.8 billion light-years – matter seems to be homogeneously distributed, even though the question of homogeneity on these scales is an active subject of modern research. As the story unfolds in this chapter, it will become clear that it was the great American observational astronomer Edwin Hubble who made emphatic statements in the 1930s that, on all scales, the galaxy distribution is homogeneous (Hubble 1936a, p. 553). In this same era, Hubble's rival, Harlow Shapley, reported what he believed to be significant irregularities in both nearby and more distant galaxy distributions. Shapley was correct, but his proof was not sufficiently strong to convince the early cosmology community to take the evidence seriously, and once Hubble's view was accepted, there was no turning back until this issue was revisited 25 years after Hubble's death when

3D maps of the galaxy distribution were made based on galaxy redshift surveys. Now, I recount the early history of the subject.

3.1   William Herschel Surveys the Sky

The gifted observational astronomer Sir William Herschel (1732–1822) was the first scientist to study the deep Universe. He did so by looking through the eyepiece of his telescope, night after dark night, identifying and, with the collaboration of his sister Caroline, cataloguing thousands of faint nebulae. Herschel and his contemporaries applied the term "nebula" (from Latin "cloud") to a diverse set of objects, some of which are local gas clouds within our Milky Way, and others that are distant galaxies. A clear distinction between the two categories was complete only when Hubble began to resolve individual stars in the nearest galaxies like the Andromeda nebula (to use its historical name).

Herschel's observational results provide a perfect "blind test" of the Cosmological Principle because he had no prejudice as to what he should expect to see. When he first catalogued the nebulae, Herschel was not even sure what these objects were, let alone how they should be distributed on the sky or in space. Being keenly aware of the crowded star fields of the Milky Way, Herschel discussed separately the nebulae that are concentrated along the band of the Milky Way and those that are scattered toward the Milky Way's northern and southern polar regions. Of course, only the nebulae seen toward the poles (those regions of the sky more than ~20° from the plane of the Milky Way) where our view is relatively unobstructed can inform the question of homogeneity of the Universe over the largest scales.

Starting in 1780, Herschel spent more than 15 years working on clear nights from his home 50 miles east of London, peering through his 20-foot telescope and calling out to his sister Caroline what he was seeing. The photographic process had not yet been invented, so these were all visual observations with verbal descriptions made, and then recorded in real time. He estimated the positions and obtained descriptions for approximately 2,400 to 2,500 nebulous objects. This manner of observing, and the effort involved by both John and Caroline Herschel, was nothing short of heroic. His son John Herschel (1792–1871) took his father's 20-foot telescope to South Africa to complete the search, and by 1820, John Herschel had catalogued another ~1,750 objects in the southern sky. Their combined effort provided the first all-sky survey of nebulae. His catalogue of nebulae is still used today and goes by the name: New General Catalogue (NGC).[1] For example, many astronomers recognize NGC 4874 and NGC 4889 as the two brightest galaxies in the core of the Coma cluster of galaxies.

Sir William Herschel (1738-1822)

William Herschel had two careers, one in classical music and the second in astronomy. He was born and educated in Hanover, Germany, migrated to England in 1757, and eventually obtained a steady position as the organist at the Octagon Church in Bath. In midlife he began to build, as a passionate hobby, his own telescopes. He first experimented with refracting telescopes but switched to building reflecting telescopes with wooden tubes and mirrors made of speculum (an alloy consisting of two-third copper, one-third tin, and a small fraction of arsenic). In this effort, he followed a text on optics written by Robert Smith in 1738. As he built his telescopes, he also started massive observational studies. While systematically searching the sky for double stars with his 6.2-inch aperture 7-foot-long telescope (he eventually catalogued 800 binary star systems), Herschel discovered the planet Uranus. For this discovery, the King of England awarded him a generous annual stipend, allowing Herschel to quit his music career and switch all his efforts to astronomy. Herschel eventually made over 400 telescope primary mirrors. His largest had a 48-inch aperture and had a focal length of 40 feet, but his most productive telescope had an aperture of 18.5-inches and a 20-foot focal length. Herschel was the first to detect binary motion in double stars – he discovered two moons of Uranus and two moons of Saturn. He was the first to see seasonal changes in the sizes of the polar caps on Mars; he was the first to detect and measure the proper motion of the Sun as it moves relative to nearby stars; and he discovered infrared radiation by passing sunlight through a prism and placing a thermometer just beyond the red portion of the visible spectrum. At age 82 (two years before his death), Herschel became the first president of the Astronomical Society of London. The image of William Herschel is a 1912 engraving after a portrait by William Artaud. Copyright The Royal Society.

Even before his own observations in the northern hemisphere were complete, William Herschel reported to the Royal Society on June 17, 1784, that he had discovered "strata of nebulae," that is, long linear structures containing hundreds of nebulae (Herschel 1784). They are located in areas away from and run perpendicular to the plane of the Milky Way. One of Herschel's strata is especially interesting; it stretches more than 90° across the sky from the constellation Centaurus to Virgo to Canes Venatici to Ursa Major. We know today that Herschel was seeing, for the first time, galaxies concentrated along the plane of the Local Supercluster. The Virgo cluster, a moderately large cluster of galaxies, sits close to the center of our Local Supercluster (at times, it is called the Virgo Supercluster), and our Milky Way galaxy lies on its outskirts. Herschel had discovered the nearest prominent structural feature in the galaxy distribution ~140 years before anyone understood the true nature of galaxies.

In 1811, when William Herschel was 71 years old, he made his eighteenth presentation to the Royal Society (Herschel 1811). It was entitled "Astronomical Observations Relating to the Construction of the Heavens … ." He described once again the constellations that are filled with nebulae, and he went one step further to describe areas of the sky (again, areas away from the plane of the Milky Way) where "the absence of nebulae is as remarkable as the great multitude of them in the first mentioned series of constellations." He listed a contiguous set of constellations relatively free of nebulae covering a large swath of sky, an area that is loosely centered on the constellation Hercules. Today, Herschel's empty region is in the same direction that astronomers have identified what is now called the "Local Void." Herschel was an astute and careful scientist with impeccable skills as an observational astronomer.

Over 100 years later, C. V. L. Charlier (1922) discussed once again the distribution on the sky of Herschel's original sample of nebulae. By this time, John Dreyer had converted Herschel's sky positions for all nebulae into the modern form familiar to astronomers today (right ascension and declination), and Dreyer renamed the end-product the New General Catalogue. The all-sky map from Charlier (1922) was reproduced in the book *Principles of Physical Cosmology* by Peebles (1993). However, Peebles did not make a clear link or connection between Charlier's all-sky map and the original data of William Herschel dating from 1811. As discussed near the end of the next chapter (in Section 4.5), Charlier himself interpreted the distribution of Herschel's nebulae in terms of a continuous clustering hierarchy, but he gained few supporters for this concept (Berendzen, Hart, & Seeley 1984).

3.2   Leavitt Prepares the Path for Shapley's and Hubble's Discoveries

In the 100 years following Herschel's great contribution, astronomers made relatively little further progress in understanding the deep Universe and the structural features in the galaxy distribution. During this time, astronomers placed greater emphasis on studying the distribution of stars around us and establishing what came to be called the "Kapteyn Universe," with an emphasis on the distribution and motion of the stars we see in our immediate vicinity. A full understanding of the Milky Way galaxy had not yet been realized, and our true place in the Universe remained beyond reach. It was in the late stages of this era when Einstein learned as a young man the little he knew about astronomy.

Before the first hints of modern cosmology began to appear, a major breakthrough in astronomy was made by the dedicated effort of Henrietta Leavitt. She worked at Harvard College Observatory for Director E.C. Pickering, where she monitored (on a series of repeated photographic plates) pulsating stars called Cepheid variables. These stars are intrinsically bright, so they can be seen over great distances. Their pulsating nature makes them stand out from their neighbors. Leavitt happened to be assigned the job of analyzing photographic plates taken in an area of the sky in the direction toward the two Magellanic Clouds. Their names are the Large and the Small Magellanic Clouds: LMC and SMC. They are satellite galaxies that orbit our Milky Way and happen to be visible from the southern hemisphere. Her job was to inspect photographs that had been taken from a telescope operated near Arequipa, Peru, by Harvard College Observatory. Leavitt would first identify and then monitor the brightness variations of the Cepheids. She was the first to recognize (Leavitt 1908) the simple fact that the Cepheids with a higher apparent brightness pulsate more slowly, and correspondingly, those Cepheids that were fainter pulsate more rapidly. Leavitt made her next big step in 1912, when she demonstrated a close mathematical relationship between pulsation period and the brightness of Cepheids (Leavitt & Pickering 1912). Once the intrinsic brightness of Cepheids had been determined by other astronomers who studied Cepheid variable stars in our Milky Way, the identification of Cepheid variable stars in distant galaxies provided the means, for the first time, to make reliable distance measurements in the local Universe.

Henrietta S. Leavitt (1868–1921)

Henrietta Leavitt was the eldest child of a Congregationalist minister and came from an affluent family. She enrolled for one year at Oberlin College in Cleveland, Ohio, but when her family moved from Ohio to Cambridge, Massachusetts, she continued her studies at what later would become Radcliff College (at the time called Harvard Annex). She graduated in 1892 and eventually became one of the many female "computers" hired by Harvard College Observatory's Director E. C. Pickering to analyze astronomical observations. One of her projects was to identify and measure the brightness of variable stars on photographic plates taken with the 24-inch Bruce astrograph located at Harvard's Boyden Station near Arequipa, Peru. In a scientific paper published in 1908, she reported the identification of 1,777 variable stars: 808 in the Large Magellanic Cloud and 969 in the Small Magellanic Cloud (Leavitt 1908). Both Magellanic Clouds are satellite galaxies of our Milky Way. Even by 1908, she had already recognized that the brighter variable stars had longer cycling periods – some on

(CONT.)

the order of months – while the fainter ones had shorter periods: on the order of days. In a follow-up paper four years later (Leavitt & Pickering 1912), she extended her earlier work and reported her discovery of the Cepheid period-luminosity relation. This set the stage for Hubble's discovery of the distance to spiral galaxies in 1923 and 1924. Hubble's distance determinations relied on Leavitt's period-luminosity relation as applied to extremely faint Cepheid variable stars that Hubble discovered in the nearest galaxies. A revolution in cosmology followed. Refs.: Berendzen, Hart & Seeley (1984), Ferris (1983, 1989), Smith (1982). Photo reproduced by permission: copyright Harvard College Observatory, Astronomical Glass Plate Collection.

Soon after Vesto Slipher (see Section 1.3) and Henrietta Leavitt made their respective breakthroughs, Harlow Shapley and Edwin Hubble began their work at Mt. Wilson Observatory. Shapley, who was four years older than Hubble, joined the Mt. Wilson Observatory staff in 1914. Hubble joined the staff in 1919. As Hubble's biographer G. Christiansen (1995) points out, they were competitors who had large egos, and their personalities seemed to have been fundamentally incompatible (see also Gingerich 1990, 1999). These points are significant because Shapley (after 1932) eventually argued in favor of striking irregularities in the local galaxy distribution, but Hubble ignored these findings and by the mid-1930s, Hubble seems to have begun basing his conclusions about the local homogeneity of the galaxy distribution primarily on matters of principle. It is of historical importance to recount what happened early in their careers in order to reveal why Hubble would ignore Shapley's later results.

Shapley had good success in measuring distances to globular star clusters. His procedure was to first identify pulsating variable stars in each cluster, and once he had measured their pulsation periods, he applied the Leavitt period-luminosity relation to obtain the intrinsic luminosities of the variable stars (Shapley 1919). Once he had both the intrinsic and the apparent luminosities, he could calculate their distances. Shapley repeated this exercise for many globular clusters. Next, he noticed that a majority of the globular star clusters were preferentially located in the direction of the constellation Sagittarius, and he then suggested that the system of globular star clusters defined the center of the Milky Way galaxy. This work was of fundamental importance, but unfortunately, Shapley, without knowing what went wrong, had mistakenly exaggerated the distances to the globular star clusters. By doing so, he deduced a huge size for the Milky Way galaxy and very quickly suggested that the Milky Way system dominates the entire Universe. It turned out that Shapley's good friend and fellow staff member at Mt. Wilson Observatory Adriaan van Maanen (1884–1946) had reported for a number of years that spiral nebulae could be seen to rotate,[2] so

Shapley incorporated this bogus information into his model and concluded that spiral nebulae were nothing more that small satellites situated in the Milky Way's halo. This was the state of affairs in 1918, just before Hubble joined the staff of Mt. Wilson Observatory. The over-sized Milky Way confused and concealed the true nature of the Universe, and it appears that Hubble, in 1919, set out on his own long-term research programs to set things straight.

In 1921, Shapley was given the opportunity to leave Mt. Wilson when he was offered the position of Director at Harvard College Observatory. This left Hubble at Mt. Wilson Observatory free to pursue his own research program on the nature of the nebulae, without any local tension with Shapley, even though fellow Mt. Wilson scientist van Maanen remained a thorn in Hubble's side for many years. Hubble's first success in rectifying Shapley's scientific errors came in 1923/1924, when Hubble definitively showed that spiral nebulae are separate islands of stars, located far outside our own galaxy.[3] He did so by following in Shapley's footsteps to identify very faint Cepheid variable stars in the handful of spiral nebulae that happen to be situated closest to our Milky Way galaxy. Hubble's discovery was a profound and revolutionary development, and it was the definitive step in establishing the structure of the Universe as we know it today. But it did not immediately resolve the issue of the over-sized Milky Way. Shapley's work suggested a diameter for the Milky Way of 300,000 light-years based on the globular cluster system. This Shapley compared to the diameter of the Andromeda galaxy (based on Hubble's new distance) of only 42,000 light-years. Shapley believed that our own local star cloud (the stars in the Milky Way as we see them in the sky) had a diameter of 6,500 light-years, while the Large Magellanic Cloud had a diameter of 11,000 light-years. To save his dominant Milky Way hypothesis, Shapley suggested, at this point, that the Milky Way system was a "Super-Galaxy" consisting of a swarm of numerous star clouds centered on the more extensive globular cluster system (Shapley 1930a).

Next, Shapley mounted a search (with Harvard College Observatory telescopes) to identify on wide-field astrographic photographs other examples of super-galaxies to bolster his case. His prime example was what he called the Coma-Virgo super-galaxy, the object we know today as the Local Supercluster (alternatively the Virgo Supercluster). His second super-galaxy example was a cloud of galaxies in Centaurus, an object known today as the Shapley Supercluster of galaxies (Shapley 1930b). By confusing our Milky Way galaxy with what modern astronomers call superclusters of galaxies, Shapley stumbled again in his effort to place an over-sized Milky Way into a comprehensive model of the Universe. His error was quickly recognized by the likes of Sir Arthur Eddington in England and, especially, Jan Oort at the Groningen Astronomical Laboratory in the Netherlands. In 1927, Oort had pioneered an analysis that made good sense of the rotational motion of stars within the Milky Way galaxy as we know it today.[4] I am unaware of Hubble's reaction to Shapley's super-galaxy hypothesis, but Hubble is not likely to have been supportive.

This is the end of the Shapley–Hubble "prelude" and explains why it was easy for Hubble to dismiss, or at least greatly discount, Shapley's subsequent work described later in this chapter on the galaxy distribution. When Hubble turned his back on Shapley's misconstrued concept of a Milky Way super-galaxy and proceeded with his own ambitious investigation of the Universe, he left it to other astronomers to figure out how Shapley went wrong. Shapley's errors were corrected in two steps. First, Shapley was unaware of problems caused by obscuring dust within our Milky Way galaxy. The difficulties caused by dust obscuration were a general problem in astronomy in this era, and by the 1930s, a better understanding was in hand. Second, Shapley had confused his favorite variable stars in globular clusters (RR Lyrae variables) with the brighter Cepheid variables seen in the younger star-forming regions of the Milky Way. Both effects made the globular clusters appear to be considerably more distant and, therefore, more widely separated from each other than they are in reality. By 1952 (the year before Hubble died), Mt. Wilson astronomer Walter Baade (1893–1960) helped to complete the job of showing that our Milky Way is an average denizen of the universe of galaxies (Baade 1951), by detecting differences between the young stellar population in the Milky Way disk and the older stellar population in globular clusters.

Both Harlow Shapley and Edwin Hubble reached the pinnacle of their respective careers in the early 1930s, when a majority of the issues just described were playing out. Hubble had the clear advantage as an observational astronomer with continuous access to the two largest telescopes in the world, both at Mt. Wilson Observatory: the 100-inch and the 60-inch aperture reflectors. Mt. Wilson, as a site for observing, was superb, although the lights of Los Angeles were a growing problem. Both Mt. Wilson telescopes had excellent instruments and support. However, the downside of the 100-inch and the 60-inch was their limited field of view. Plates from the 100-inch telescope captured less than 0.5 square degrees of sky, and those from the 60-inch captured less than 0.75 square degrees.[5] From Harvard, Shapley had access to two smaller astrographic telescopes capable of taking images that recorded 9 square degrees at a time. Harvard maintained two astrographs: one in the northern and the other in the southern hemisphere. But these smaller instruments both had apertures of 24 inches, so they could not explore to the same depth as the 60-inch and 100-inch telescopes. Both astronomers took a great interest in whether or not the galaxy distribution is homogeneous.

3.3   Shapley Identifies Local Inhomogeneity

By the early 1930s, Shapley had fully accepted the extragalactic nature of spiral nebulae, and he and his assistant Adelaide Ames (1900–1932) began to catalog and study the distribution of the brightest galaxies in the sky. These they identified from a complete set of astrographic photographs taken over both the northern and southern galactic polar caps. Their all-sky survey of the brightest (and generally the nearest) galaxies was published in 1932 (Shapley & Ames 1932a, 1932b). Their catalog is somewhat smaller but closely resembles the survey of nebulae published more than 100 years earlier by William Herschel, and it repeats a more limited study in 1923 by John Reynolds (1874–1949). Hubble's discussion of Shapley's work makes it clear that other astronomers in the 1930s were well aware of the structural distribution of galaxies in the northern constellations that Herschel first identified. Shapley and Ames point out that Herschel had already identified 91% of their bright galaxy sample. Because of the significant overlap, it comes as no surprise that Shapley and Ames describe essentially the same features first recognized by Herschel.

Harlow Shapley (1885-1972)

 Harlow Shapley was born on a farm in Missouri. He completed elementary school but went no further at that point and became a news and crime reporter. After studying at home through his teens, Shapley returned to high school at age 19 and within 18 months graduated as Valedictorian of his class. Shapley earned a B. A. and M.A. at the University of Missouri under Fredrick Seares (who later moved to Mt. Wilson Observatory) and a PhD at Princeton University under Henry Norris Russell. In 1914, at age 29, Shapley joined Mt. Wilson Observatory as a staff astronomer. He learned from Russell how to apply the Leavitt period–luminosity relation to measure astronomical distances after systematically identifying variable stars in numerous globular clusters. In 1909, Karl Bohlin had suggested that globular clusters formed a coherent system located at the center of the Milky Way, a hypothesis that Shapley and others rejected at first, but by 1918, Shapley fully embraced this concept. Shapley used his distances to globular clusters (uncorrected for dust extinction) to determine the size of the Milky Way galaxy. He was the first to propose that Cepheid variable stars pulsate thereby causing their light variation. Shapley also discovered a new type of galaxy called "dwarfs" by identifying and measuring the distances to the

In Section 7 of the Shapley–Ames Catalogue, in a section entitled "Distribution in the Sky," Shapley and Ames make two key points based on the all-sky map shown in Figure 3.1. First, in the northern galactic hemisphere, there are twice as many bright galaxies as there are in the southern hemisphere. Second, there are conspicuous vacant regions in both hemispheres where bright galaxies are absent. The first point is important because the north–south asymmetry was referenced repeatedly in Shapley's later research on the galaxy distribution. Shapley and Ames elaborate on the second point in their supplementary paper (not the catalogue itself), where they say that the vacancies "are not due to the insufficiency of the survey … nor are the barren regions, such as that at $\lambda = 20°$, $\beta = +50°$, the result of heavy obscuration." The location $\lambda = 20°$, $\beta = +50°$ is centered in the constellation of Hercules, where Herschel also saw a deficiency of nebulae. The 1932 analysis is a modern re-discussion of Herschel's results from 1784 and 1811, and it fit nicely into Shapley's super-galaxy hypothesis in which an extended collection of star clouds formed a distinct system, outside of which the number of star clouds diminished. As best I can tell, Shapley never mentioned the vacant regions in any other publication after 1932, although he did repeatedly discuss the north–south asymmetry of the galaxy distribution, clearly revealed in the Shapley–Ames Catalogue.

In 1934 and 1938, Shapley discussed the large-scale distribution of faint galaxies (as seen on photographic plates taken with the Harvard astrographs) in terms of what he called the "metagalactic density gradient."[6] His concept developed as follows: since the Shapley–Ames Catalogue identified a factor of two difference in the galaxy density between the northern and southern hemispheres, there must be cases where a similar density contrast can be seen in counts of galaxies projected onto the sky. Shapley then searched for and successfully found significant gradients in the surface density of galaxies. His 1934 paper on this phenomenon was entitled *A First Search for a Metagalactic Gradient* and his 1938 paper *A Metagalactic Density Gradient* (Shapley 1934, 1938). Shapley emphasized that he was searching for galaxy surface gradients as projected onto the sky, and there was no effort placed on identifying specific regions that were devoid of galaxies.

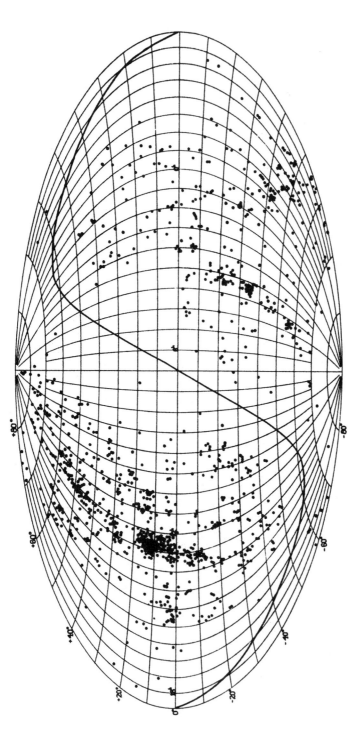

Figure 3.1 Shapley-Ames all-sky distribution of bright galaxies. Each point shows the location on the sky of a nearby bright galaxy. The S-shaped solid line running across the center of the diagram represents the plane of the Milky Way. Near this line, distant galaxies are hidden behind dust clouds in the Milky Way. The area on the upper left is the north galactic pole region, and the area on the lower right is the south galactic pole region. This diagram shows the remarkable local asymmetry of the galaxy distribution between the north and the south. The dark clump of points on the left is the core of the Virgo cluster of galaxies. Extending above and below it are galaxies associated with the plane of our Local Supercluster. With permission: copyright Harvard College Observatory.

3.4   Edwin Hubble Explores the Universe

Edwin Hubble charted his own course in his investigation of the galaxy distribution and rarely commented on the density gradients that were so clearly described by his contemporary Harlow Shapley.[7] Hubble championed a homogeneous Universe filled uniformly with "field" galaxies. He acknowledged a single type of structure within the galaxy distribution: isolated clusters of galaxies scattered here and there that contained less than a few percent of the total galaxy population. Hubble's concept was that clusters were embedded in the background of "field" galaxies. He stated clearly that, in his view, clusters were not significant on the large scale because the galaxy distribution tended toward homogeneity.[8] Hubble's conclusions were drawn from his monumental study of the deep Universe based on more than 2,000 photographic images he took with the Mt. Wilson telescopes. Hubble began this program in 1926, shortly after establishing the distances to spiral nebulae. The most significant results were published in two papers in the *Astrophysical Journal* (Hubble 1934, 1936a).

Hubble's strategy was to take many very deep images, each with a small field of view, on a set of grid points that were widely spaced across the portion of the sky that was visible from Mt. Wilson Observatory, see Figure 3.2. His aim was to test for global isotropy of the Universe by counting all galaxies, especially the faintest and, therefore, the most distant galaxies that could be detected. Hubble found that each small survey photograph he took recorded an average of ~100 galaxies that he could clearly identify (and many fainter galaxies that were just beyond the reach of his telescope).

Hubble's observational program might be compared to fishing in the Earth's ocean. He dropped fishing lines on a grid with a spacing of 500 miles, caught about 100 fish with each line (some large and some small) in order to test the contents of the oceans. He found a relatively consistent population from one test point to another, but he had no information as to how the fish might be organized across the globe into schools or groups at any spacing smaller than his grid-point separation, which, in actuality, ranged from 5° to 10° on the sky.

Hubble's first paper on this topic, published in 1934 in the *Astrophysical Journal*, contains his quantitative results showing the number of galaxies detected at all grid points. Away from the plane of the Milky Way, he found an average of 100 galaxies per field. Some fields contained fewer than 100, while others had as many as 350. Hubble analyzed the sample near the poles of the Milky Way (away from the dusty obscuration in the plane of our galaxy) and plotted a graph that showed how many fields (how many photographs) had zero galaxies, 10 galaxies, 20 galaxies . . . up to his maximum limit. He found a bell (i.e., Gaussian) curve when he plotted not the straight counts of galaxies but a logarithm of the

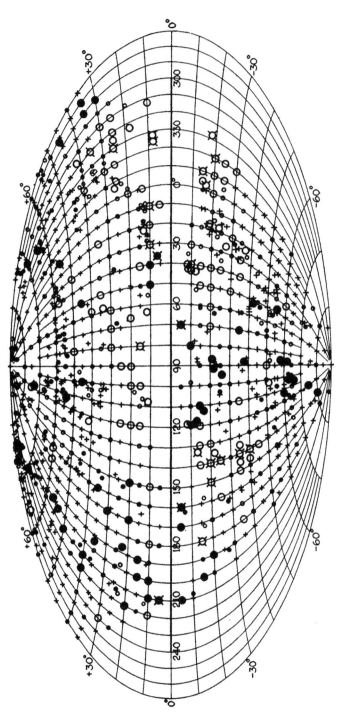

Figure 3.2 Hubble all-sky faint galaxy counts. Hubble counted galaxies in survey fields distributed widely across the sky on a grid with an approximate 5° spacing. This equal-area Aitoff projection shows the positions of his fields in an all-sky map with a coordinate system defined by the plane of the Milky Way. Along this plane (the central horizontal line in the plot above), the obscuration by dust in the Milky Way is confined to a broad central "equatorial" band, while there are perfectly clear areas high in the northern and southern galactic hemispheres. It was in these galactic polar regions that faint galaxy counts are most dependable. By permission of the A.A.S.: E. Hubble (1934). *Astrophys. J.*, 79, pp. 8–76.

counts. Mathematically, this is called "log-normal." Hubble took this finding as an indication that the galaxy population is homogeneous in depth.

Edwin P. Hubble (1889–1953)

Edwin Hubble accepted a prestigious staff position at Mt. Wilson Observatory in 1919, five years after Shapley arrived. In 1919, Shapley was at a high point in his career, having just proposed a revolutionary (but ultimately incorrect) model of the Universe containing a dominant Milky Way galaxy. From the start, Hubble projected himself as aloof and arrogant and consistently used a feigned British accent. Hubble was the first astronomer to carefully and systematically measure distances to external galaxies in the local Universe. This success came in 1923/1924 when he used Cepheid variable stars to obtain distances to the nearest neighbors of the Milky Way. In subsequent years, Hubble cautiously extended his range to greater and greater distances. Then in 1929/1930, with the help of his assistant Milton Humason, Hubble made a great leap forward by confirming the linear relation between the apparent Doppler shift ("velocity") and distance for galaxies, a relationship that (when plotted) was called the "Hubble diagram." By doing so, Hubble placed our Milky Way galaxy in its proper setting in the Universe as a whole. Hubble's work reached its pinnacle in 1936 when he published a technical but semipopular book entitled *The Realm of the Nebulae* based on lectures he had given at Yale University. Hubble advocated a simple model for the galaxy distribution: a smooth background of "field galaxies" with occasional rich clusters scattered at wide intervals. The rich clusters were estimated to contain perhaps 10% of the total galaxy population and were not viewed as significant perturbations to what he believed to be a homogeneous Universe. Primary source: Christenson (1995). Image courtesy of the Edwin Powell Hubble Collection (1033–5) at The Huntington Library, San Marino, California.

We know today that the highly structured 3D distributions (i.e., what we call today the cosmic web) show log-normal characteristics when projected galaxy counts are inspected in fixed "boxes." It is ironic that a highly structured galaxy distribution could look so simple and benign when projected in 2D on the sky. Hubble even revealed the weakness in his analysis technique when he introduced and briefly discussed (in his 1934 paper) the Shapley–Ames galaxy sample. Recall from the earlier text that Shapley and Ames found significant inhomogeneities in the local galaxy distribution, because the sample contains the Local Supercluster and sits at the edge of what we call today the "Local Void." In his 1934 paper, Hubble followed his standard recipe and counted the Shapley and Ames galaxies in boxes on the sky, after which he plotted these bright galaxy counts in the same manner as his own faint galaxy sample. Hubble then reported that the Shapley–Ames galaxy sample also displays the same log-normal characteristic, similar to his all-sky survey, despite the fact that this particular sample shows dramatic irregularities on the plane of the sky! It is not known whether Hubble was puzzled by the fact that the local inhomogeneous galaxy distribution displayed log-normal characteristics too.

Throughout the decade of the 1930s, while Shapley never directly confronted Hubble on the subject of the homogeneity of the galaxy distribution, Shapley was not hesitant to present his own views. As described in detail by Gingerich (1990), at the 1932 meeting of the International Astronomy Union (which Hubble did not attend), Shapley made his case for significant irregularities in the galaxy distribution to the luminaries of the cosmology community, including Arthur Eddington and George Lemaître. But from a historical perspective, Shapley's work seems to have had little to no impact whatsoever.[9] Furthermore, Bart Bok, who was a Harvard professor in 1934, wrote a brief commentary on the nature of the galaxy distribution (Bok 1934) based on data from Shapley and Ames (1934a) as well as from Hubble (1934). Bok showed statistically how both data sets revealed evidence for significant clustering. But just like the treatment given to Shapley, Bok's work was largely ignored at the time and has rarely been referenced.

Hubble's second major publication on the galaxy distribution came in 1936 when he presented observations to test whether the Universe is homogeneous in depth. To do so, he started with the highest-quality one-hour exposures from his 1934 paper (consisting of 214 photographic plates from the 60-inch telescope and 228 from the 100-inch telescope), and supplemented them with 121 plates from the 60-inch (each a 20-minute exposure) and 41 excellent two-hour exposures from the 100-inch telescope. Then, he incorporated 284 survey plates (each a one-hour exposure) taken by Nicholas Mayall of Lick Observatory, who had used the Crossley 36-inch telescope. Hubble counted the number of galaxies on each photograph, averaged these numbers for each of these five surveys, and plotted them as a function of the limiting magnitude. His key result is shown in Figure 3.3.

There were two significant conclusions that Hubble drew from his analysis of these new data. First, the positive news: the galaxy distribution to the faintest limits available in his two-hour exposures with the 100-inch telescope shows no indication of ending. That is, the number of galaxies continues to rise in

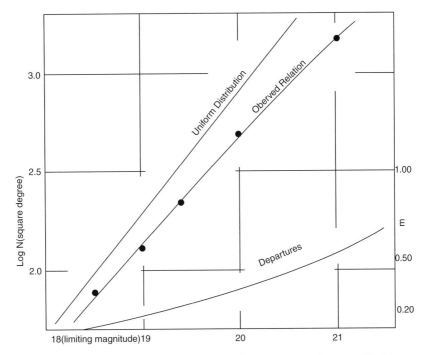

Figure 3.3 Hubble's 1936 galaxy counts. Galaxy counts are shown as filled circles, and these he compared to a standard model Universe with uniform density. The lower curve shows the systematic departures. The text explains how Hubble's interpretation of these departures led him astray. By permission of the A.A.S.: E. Hubble (1936a). *Astrophys.* J., 84, pp. 517–54.

a monotonic fashion as the samples get fainter and fainter. Astronomers as early as William Herschel had performed similar fundamental tests using counts of stars. In the tests with stars, the outer edge of our Milky Way galaxy was detected because the number of stars is definitely limited (in particular, when viewed in the direction perpendicular to the plane of the Milky Way). But Hubble saw no limit to the observed galaxy distribution in any direction in the north and south galactic polar regions. From this result, Hubble drew the extremely profound conclusion that galaxies were reliable "markers in space" and that our local neighborhood is but a small fraction of a very extensive homogeneous universe.

Although Hubble estimated that his survey extended to a redshift of $z = 0.2$, a modern reassessment shows that Hubble's two-hour exposures with the 100-inch telescope reached beyond this to $z \sim 0.4$. He was, indeed, sampling the deeper Universe, and in this realm, modern investigations reveal that evolutionary corrections are required to properly interpret deep galaxy counts: stars in the distant galaxy sample are younger than stars in more local galaxies. Yet Hubble argued in his 1936 paper that evolutionary effects were likely to be small. This incorrect assertion, as well as his low value of $z = 0.2$ for the estimated limit of his deep survey, led Hubble into an inescapable bind.

In his 1936 paper, Hubble presented two models of the Universe in an effort to explain his galaxy counts. The first was an insightful harbinger of the future. He had collaborated in this particular analysis with his Caltech colleague Richard Tolman, starting in 1934, to develop the tools to match his data to a standard general relativistic expanding Universe. While Hubble was making a bold step into modern cosmology, the fit Hubble found to his faint galaxy counts was a Universe that was small, rapidly expanding, had a high mean density and a cosmological constant. In the concluding paragraph of his 1936 paper, Hubble calls this high density Universe "disturbing." Therefore, he rejected this model even while presenting it. Hubble's second model was his favorite, but in the context of modern astrophysics, it is equally disturbing. The model was a throwback to the days before Einstein in the sense that the observed galaxy redshifts were required to be not a result of any velocity, neither conventional velocities nor induced by a general relativistic expansion of space. Hubble insisted that the galaxy redshifts could, instead, be caused by an as-yet-unexplained physical process acting uniformly throughout the Universe.[10] In this second model, Hubble postulated a Universe that was static, infinite, and homogeneous! Although the model is reminiscent of one that even Isaac Newton might have imagined, Hubble stated that it need not be entirely classical. It could be based equally on general relativity, but in this case the spatial curvature was flat and the rate of expansion inappreciable. With these radically incorrect assumptions, Hubble was able to explain his faint galaxy counts. Hubble's result stands by itself as total irony where he rejects the modern relativistic expanding universe (complete with a cosmological constant) in favor of a static infinite classical model that left the nature of the redshift a mystery yet to be resolved. Instead, what was truly missing from Hubble's analysis was an understanding of the evolutionary corrections to the galaxy magnitudes as well as a better understanding as to how deep his survey had penetrated into the Universe (Sandage 1987, 1989). Edwin Hubble would remain trapped in what modern cosmology would call a mental cul-de-sac for the rest of his life, believing that the Universe was infinite and homogeneous with the redshifts of galaxies produced by an as-yet unexplained physical process.

Milton L. Humason (1891–1972)

In 1919, when Edwin Hubble arrived at Mt. Wilson Observatory, Milton Humason was promoted to be an observatory staff scientist. Humason was originally hired as a janitor, but soon volunteered to run the telescopes as a night assistant. He had had no formal education and approached his work in a simplistic manner. He had excellent technical skills, a smooth personal demeanor, and consistently behaved like a true gentleman. By the late 1920s, Humason was both exposing through the night and measuring spectra of faint galaxies with the 100-inch telescope for Hubble. Humason's observations were used to confirm the linear nature of the velocity–distance relation, a relation that sits at the foundation of modern cosmology. Edwin Hubble's biographer, Gale Christianson, recorded a story told by Milton Humason that might have triggered Hubble's further distrust of Harlow Shapley. This is the story. Immediately before Shapley departed Pasadena for Harvard in March 1921, Shapley gave to Humason photographic plates of the Andromeda nebula (M31) for examination with the blink stereo comparator (the same machine used by van Maanen to measure bogus rotational motions of nebulae). It allows the astronomer to inspect under magnification two images in quick succession by blinking from one to the other. During Humason's work, he came upon what appeared to be very faint pulsating variable stars in Andromeda (M31). Humason marked them on the photographic plates with an ink pen (writing on the opposite side from the emulsion does not affect the recorded image) and showed them to Shapley. Given the very faint appearance of these stars, if these were variables of the same nature as those Shapley studied in globular clusters, it would place the Andromeda nebula far outside the Milky Way galaxy and contradict Shapley's model of the Universe. According to Humason, Shapley "was having nothing of this." Christianson writes, "then [Shapley] calmly took out his handkerchief, turned the plates over, and wiped them clean of Humason's marks." As reported by Smith (1982, p. 144, footnote 122), Owen Gingerich once asked Shapley about this story, and Shapley thought it might be true. Four years after the interaction between Humason and Shapley, it was Edwin Hubble who was given credit for discovering faint Cepheid variable stars in the Andromeda galaxy. Primary source: Christenson (1995). Image courtesy of the Observatories of the Carnegie Institution for Science Collection at The Huntington Library, San Marino, California.

Through the early 1940s, Hubble published only a handful of minor scientific papers. His work in astronomy had been interrupted by a commitment to public service during World War II. This alternate work largely filled the time while he awaited the completion of the 200-inch telescope. In 1949, Hubble returned to the problem of the faint galaxy counts and enlisted the help of Caltech graduate student Allan Sandage (1926–2010) who, under Hubble's direction, began galaxy counts on photographs from the new 48-inch Schmidt telescope at Mt. Palomar Observatory. Sandage's effort stalled when Hubble suffered a massive heart attack in July 1949. Hubble took time away from his work to recuperate before he began to use the 200-inch telescope in its early commissioning runs. Despite the fact that Hubble had defined and was moving forward with an extensive plan to ambitiously resurrect all of his projects from the 1930s, in the fall of 1953, Hubble died quite suddenly at age 63.

I discussed this history with Allan Sandage in 1986, during a visit he made to the Institute for Astronomy at the University of Hawaii. Sandage told me it was fortunate Hubble had not invested additional time in extending the faint galaxy count program with the 200-inch telescope. In 1961, Sandage had thoroughly reviewed galaxy counts as a means to select among what Sandage called "world models" and found such serious deficiencies in the faint galaxy count method that he called it in 1986 a "waste of time." In a prescient statement in Sandage's 1961 paper, he expressed further concern about the galaxy count method over the fact that "it is well known that galaxies are not homogeneously distributed on the plane of the sky but show a strong tendency to cluster" (Sandage 1961). This was yet another reason Sandage cited as a ground for avoiding galaxy counts for cosmological analysis.

3.5   Early Theories of Galaxy Formation

The discussion in this chapter on the galaxy distribution circa 1930 would be incomplete, from a modern perspective, without a description as to how galaxy formation and galaxy cluster formation were perceived during the early years of the twentieth century. Today, these topics are an intrinsic component of the cosmological model because the growth of galaxies and larger structures comes part and parcel with the evolution of the Universe. Yet in the 1930s, galaxy formation was seen as a separate discipline and was given its own name: cosmogony.

As it turns out, in the same era when Hubble and Shapley were investigating the nature and distribution of nebulae, Sir James Jeans (1877–1946) was working to update and extend what was then the 100-year-old Laplace nebular hypothesis. Laplace had introduced the general concept in 1796 that a contracting rotating gas cloud might be the precursor of our Solar System and that this cloud would evolve into a flattened disk of planets. As the initially spherical gas cloud contracts, it flattens; a disk forms, and the disk fragments into planets. Jeans began a serious effort in the early 1900s to place the Laplace concept on a solid foundation of mathematical physics, and in his honor, the term "Jeans instability" is used today to describe the slow contraction of material in a slightly over-dense region of space. Jeans developed these ideas for two different initial states of matter: a contracting spherical gas cloud as well as a spherical collection of stars, both with initial overall rotation. By 1919, the year Hubble started to work at Mt. Wilson Observatory, Jeans had already made good progress (Jeans 1919).

According to Hubble's biographer, Gale Christianson (1995), Jeans and his family visited Hubble and his wife in California twice, once in 1923 and a second time in 1925. This was the period during which Hubble was resolving the nature of the so-called spiral nebulae. When it came time for Hubble to devise a system to place galaxies in categories of similar appearance (astronomers call them morphological classes), these categories as well as the order in which Hubble placed them, had a striking resemblance to the expected evolutionary processes that Jeans had studied. Hubble was somewhat brash to give labels to these categories when he called the smooth spherical ones "early" and the flattened ones with abundant structure "late," making it seem that the morphological sequence of galaxies was also an evolutionary sequence. It was brash because there was no evidence to support the assertion that the morphological sequence was also an evolutionary one. (An alternate view is given by Baldry 2008.) In fact, Hubble had the evolution reversed in the sense that his "early" types like elliptical galaxies are known today to be the most ancient while his "late" types like spiral galaxies are still forming young stars.

Despite criticism by some astronomers, Hubble's labels of early and late (along with their evolutionary implications) "stuck" and Hubble's morphological sequence and Jeans' theoretical concepts formed the basis for the first model of galaxy formation. The extremely premature nature of these concepts of galaxy evolution must be emphasized. It was in this same period that Hubble's adversary at Mt. Wilson Observatory van Maanen was confusing everyone by claiming to see spiral nebulae rotate. In his early scientific publications, Jeans considers van Maanen's work to be statements of fact that he tries to fit into the constraints of the evolutionary model.

There were also attempts in the 1930s to understand the origin of galaxy clusters, and among the scientists who participated are two Swedish astronomers, Knut Lundmark (1889–1958) and Erik Holmberg (1908–2000). I quote here from a paper by Holmberg (1937) in which he states:

> K. Lundmark in several papers discusses the origin of double and multiple galaxies. The frequency of captures between the components

of the metagalactic system is found to be very great, one every 3500 years. Thus, we ought to have a great many physical double systems. These will then act as condensation nuclei attracting new members. In this manner the systems will grow larger and larger and the possibilities for the formation of groups and clusters are given.

Therefore, in the 1930s, astronomers were beginning to develop a picture of hierarchical growth of galaxy clusters from what Hubble called the "general field of galaxies." In this same paper, Holmberg describes how the effects of strong tidal interactions between galaxies affect the rate of growth of binary galaxies. For an example of one such interaction, see Figure 3.4. Holmberg went further and used the number of binary and multiple galaxies to calculate the age of the Universe. The details were incorrect because the early distance scale (directly tied to the "Hubble constant") was off by a factor of ~10. The underlying assumptions, however, are based on Hubble's view that galaxies were initially distributed homogeneously in space.

Figure 3.4 Whirlpool Nebula M51. Discovered in 1776 by Charles Messier, M51 is a classic example of a pair of tidally interacting galaxies. The small companion sits somewhat behind the larger galaxy: clouds of obscuring dust in a spiral arm can be seen projected over the central body of the companion. M51 is a near-neighbor of the Milky Way, being at a distance of 23 million light-years. (This makes it only 10 times further away than the nearest giant spiral, Andromeda.) One-hour exposure taken by the author in September 1981 at the prime focus camera of the Canada-France-Hawaii Telescope (Mauna Kea Observatory, Hawaii). With permission: copyright Canada-France-Hawaii Telescope Corporation.

The scientific work on the large-scale galaxy distribution of both Edwin Hubble and Harlow Shapley lost momentum during the years of World War II. Shapley was more successful in reviving his work after the war. He subsequently published another ~100 research articles as well as a book before his death at age 86 in 1972. Among this late-life work, the most interesting was his 1958 book in which he discusses the need to use newer wide field-of-view cameras that are free of "edge distortion" to study the distribution of galaxies. More than anyone else, Shapley was aware of the nonuniform nature of the images produced by the two Harvard astrographs: at the outer perimeter, the images were poorly defined. He wondered how much of an effect these non-uniformities contributed to his conclusion that the local galaxy distribution is inhomogeneous. Knowledge of this deficiency of Shapley's telescopes may have influenced Hubble when he chose to ignore the significance of Shapley's work.

4

All-Sky Surveys in the Transition Years 1950–1975

Hubble established in the mid-1930s the general homogeneity and isotropy of the Universe. He used the largest telescope in the world at that time to probe to the greatest depths possible with photographic plates, and the main weakness was his sparse sampling on the sky: he used photographs with very small fields of view on a grid of widely separated points across the sky. On the other hand, Harlow Shapley reported irregularities in the distribution of bright nearby galaxies in larger fields of view (confirming Herschel's visual findings), but he was unable to convince the cosmology community that the galaxy irregularities he detected were especially meaningful. This situation left fertile ground for C. David Shane (1895–1983) who was appointed Director of Lick Observatory in 1945, immediately after World War II. Shane put the finishing touches on a new 20-inch double astrograph telescope originally funded 10 years earlier for another project. In 1947, Shane began a photographic survey that eventually captured images of the entire northern sky in exposures, six degrees on each side. By spacing his field centers on a five-degree grid, adjacent survey images generously overlapped, so it was possible for him to extend a uniform brightness calibration from one photographic plate to the next across the sky. The Lick astrographs, unlike Shapley's telescopes, maintained good image quality to the edges of each photograph. For the first time, it was possible to reliably study the faint galaxy distribution over large angular areas on the sky.

Simultaneously, a second highly significant all-sky photographic survey was underway at Mt. Palomar Observatory. While astrographic cameras, like those used by Shapley and Shane, might lose some image clarity near the edges of their fields of view, the new telescope at Mt. Palomar was based on a Schmidt optical design. It produced images with near-perfect definition over the full square field of view that was six degrees on each side. The Mt. Palomar telescope had a larger aperture of 48 inches (the Harvard and Lick Observatory telescopes had apertures of

24 inches and 20 inches, respectively), so it peered deeper into the Universe. Managed by Ira Bowen, Director of the Mt. Wilson and Mt. Palomar Observatories, and with survey funding from the National Geographic Society, this survey was completed in 1956. Shortly thereafter, copies of the National Geographic Palomar Observatory Sky Survey were offered to observatories worldwide.

In every way the times were changing. Two sky surveys were underway, and a new wave of investigators was replacing the older generation. Harlow Shapley retired as the Director of Harvard College Observatory in 1952, and Edwin Hubble died in late 1953. At about this same time, Shane stepped forward to carry out the Lick astrographic survey, and a Caltech graduate student named George Abell (1927–1983) began working as a team member of the Palomar Sky Survey. Abell's lifelong contributions, both to astronomy in general and to the study of rich galaxy clusters in particular, would be highly significant. Gerard de Vaucouleurs (1918–1995) had moved with his wife Antoinette de Vaucouleurs (1921–1987) from France to England to Australia and then to the United States. He eventually accepted a position as professor of astronomy at the University of Texas in Austin. De Vaucouleurs would advance our understanding of the Local Supercluster far beyond the work of Herschel, Shapley, and Ames by mapping its 3D structure with his own estimates of galaxy distances. Also working throughout this era was Fritz Zwicky (1898–1974), a close associate of Edwin Hubble. While Zwicky made remarkably creative – even brilliant – contributions to astronomy, ironically he held fast to Hubble's antiquated views on the galaxy distribution and denied the existence of any significant large-scale structure in the galaxy distribution. Finally, there were intellectual naysayers primarily from Princeton University who, despite all evidence to the contrary, insisted that the local galaxy distribution was likely to be homogeneous. Along with Zwicky, those from Princeton held out the longest for the old traditional views. Historians of science surely will find their perspective interesting. Those of us who first discovered cosmic voids found these traditional views to be exasperating.

4.1   Shane and Wirtanen Sky Survey

C.D. Shane's survey from Lick Observatory was the first to bear fruit. The earliest maps of the galaxy distribution, from just one section of the sky, were released by Shane and his collaborator Carl Wirtanen (1910–1990) in 1948, with a revised version published in 1954. Their publications contained figures showing maps of the sky with smoothed contours somewhat resembling topographic maps of mountains and valleys on Earth (Shane & Wirtanen 1948, 1954).

The contour maps revealed numerous regions of high galaxy density sitting above the general background galaxy population; of course, the peaks were galaxy clusters. Also visible were extended "clouds" of galaxies – what would be called "superclusters" today – but much of the cloud-like structure was neither clear-cut nor distinct. Shane and Wirtanen's survey spanned 70% of the sky (all that was visible from Lick Observatory), and in total they detected 800,000 galaxies. Their results were recorded as galaxy counts in boxes 10 by 10 arcmin square. For the early 1950s, the sample size was huge.

Shane was quick and generous to share these data with other astronomers, and because of the large size of the data set, he arranged a collaboration with Jerzy Neyman (1894–1981) and Elizabeth Scott (1917–1988), both of whom were professors of mathematics at the University of California, Berkeley. Neyman had come to Berkeley in 1938, and by 1955 he had established a separate program at Berkeley in mathematical statistics. Scott was a professor of mathematics who worked throughout her life in astronomy. Berkeley statisticians hired (human) calculators to assist in the analysis.

From a modern perspective, their most interesting paper was Neyman and Scott (1952). It set forth a model of the galaxy distribution in which every galaxy was hypothesized to be, in a formal statistical sense, a member of a cluster. Recall that Hubble and Zwicky believed that ~90% of the galaxies were relatively isolated "field" galaxies, and to accommodate this idea, Neyman and Scott acknowledged that there may be "clusters" consisting of a single member. Then, in a joint publication with Shane, they fit their model to a portion of the Lick galaxy counts and showed that by adjusting three flexible parameters, their model could easily match the actual galaxy counts on the sky (Neyman, Scott, & Shane 1953). For their best model, the three adjustable parameters fell in a reasonable range. In one of these papers, they reversed the process by making a synthetic field of galaxies with fake objects randomly placed in a way that was consistent with the three adjustable parameters of their model. When judged visually, it qualitatively matched what was observed on the sky (Scott, Shane, & Swanson 1954). Today, it is no surprise that such a model would fit the observations. We know that galaxies belong to (are confined within) the cosmic web, and when projected onto the sky, it appears as though galaxies are all associated with small groups and clusters. If Neyman and Scott had been prescient and had also incorporated large empty voids (with adjustable dimensions) in their model, no doubt it would have fit the observations at least as well, if not better.

Neyman and Scott found many other applications for the Shane and Wirtanen galaxy count data and eventually published more than a dozen other papers on various aspects of the galaxy distribution. Shane himself did the same, and in October 1970, he completed an invited review chapter in the widely read book *Galaxies and the Universe* on his general findings (Shane 1970). In his concluding remarks, Shane states that the Lick galaxy counts show wide array of structure – over scales of up to 100 million light-years (30 Mpc) – with "suggestive evidence of still larger assemblages of galaxies on a scale" of 300 million light-years (100 Mpc) or more. Shane even states: "There may be no important background of field galaxies," and he clearly describes a list of superclusters that I have included in Table 4.1. Neyman, Scott, and Shane's results are perfectly consistent with what we know today about the cosmic web, and even though these research results were widely circulated and discussed in the 1950s, they were not considered in any significant manner by cosmologists who stuck with homogeneous models.

A completely separate study of the Shane and Wirtanen galaxy counts was made by D. Nelson Limber (1928–1978). He worked as a graduate student at the University of Chicago's Yerkes Observatory in the early 1950s and did his work at the suggestion of and under the direction of Yerkes Observatory astronomer Subrahmanyan Chandrasekhar (1910–1995). Chandrasekhar and his Yerkes colleague Guido Munch had written papers showing how random dust clouds in our own galaxy might cause the apparent brightness fluctuations seen along the Milky Way (Chandrasekhar & Munch 1952). Those who studied the apparent fluctuations in the numbers of distant and faint galaxies (as seen projected onto the sky) also worried about whether similar dust clouds might make the faint galaxy counts fluctuate, and Limber addressed this issue too (Limber 1953, 1954, 1957). Chandrasekhar and Munch employed a statistical method called the correlation function, and Limber applied it to the faint galaxy counts. (In this context, autocorrelation can be substituted for correlation when individual galaxies in a survey are used as reference points for the analysis.) This function is defined as the probability of finding a second galaxy within a specified distance of a pre-selected "first" galaxy. An accurate correlation function is obtained by averaging over a suitably large sample of galaxies. This function is significant because it provides a simple statistically determined measure of structure within the galaxy distribution. As the measurement is made from galaxies that are projected onto the plane of the sky, these results obscure (are insensitive to) depth information. It turns out that simultaneously (or nearly so) Vera Rubin was also applying a similar correlation function analysis not to Shane and Wirtanen data but to Shapley's faint galaxy counts (Rubin 1954). Both Limber and Rubin were doing similar PhD thesis research. Limber's advisor was Chandrasekhar, while Rubin's advisor was George Gamow.

Many years later, Princeton University Prof. P. J. E. Peebles (b. 1935) and his students became specialists in measuring the galaxy correlation function for many different galaxy and galaxy cluster samples. The correlation function is extensively described in Peebles' 1980 book entitled *The Large Scale Structure of the Universe*.

4.2   National Geographic Palomar Observatory Sky Survey and Abell Clusters

Abell worked on the National Geographic Palomar Observatory Sky Survey while completing his PhD at Caltech. He was employed as a graduate student to take survey photographs and to inspect their quality immediately after they were obtained; those photographic plates that were inadequate would be repeated. During the initial plate inspection and more carefully afterward, Abell identified thousands of rich galaxy clusters on the sky survey photographs. His PhD thesis consisted of a statistically complete rich galaxy cluster list and a preliminary analysis of the resulting catalogue; his major results can be found in Abell (1958).

Abell's procedures were systematic and uniformly applied. Once a new cluster was found, he would identify the 10th brightest galaxy in the cluster. Its brightness was used to estimate the cluster distance. Based on this distance, he counted the number of cluster galaxies in a circular area, adjusted to the estimated distance (making it a fixed metric area). After correcting the number of cluster members for a local background count, he recorded the cluster "richness." Because the National Geographic Palomar Observatory Sky Survey images were superb and the exposures were deep, his cluster catalogue probed far into the Universe, with some clusters detected to a redshift of $z \sim 0.4$. Abell's catalog for the northern sky was such a valuable resource that when the southern sky survey photographs were completed (from Siding Springs, Australia), George Abell directed the initial effort to extend it to the southern sky. After Abell's premature death in 1983 at age 56, Harold G. Corwin Jr. and the late Ronald Olowin completed the southern cluster catalog. Combined, the northern and southern hemisphere catalogs contain just over 4,000 rich galaxy clusters.

While Abell's work with the National Geographic Palomar Observatory Sky Survey was underway, Shane and his collaborators were finishing their analysis of the Lick survey. Shane had identified galaxy superclusters (he called them "clouds") in the Lick galaxy counts. Naturally, Abell looked for the same super-cluster phenomenon in his rich cluster sample, and those he identified are listed with others in Table 4.1. After his catalogue was completed Abell compiled counts-in-cells centered on the individual clusters (looking for adjacent clusters). Abell found the counts-in-cells to deviate strongly from a random

Table 4.1 *A complete list of known superclusters as of 1970*

Name	RA (1950)	DEC (1950)	Radial velocity (km s^{-1})	Reference
Abell SC-1	00 26 00	−11 30	60,200	Abell (1961)
Perseus-Pisces	03 15 00	+41 19	5,370	Zwicky (1957)
Abell SC-2	09 18 00	+78 00	53,700	Abell (1961)
Abell SC-3	10 16 00	+50 30	49,000	Abell (1961)
Abell SC-4	11 39 00	+54 30	21,400	Abell (1961)
Abell SC-5	11 40 00	+12 00	44,600	Abell (1961)
Abell SC-6	11 45 00	+29 30	8,700	Abell (1961)
Local Supercluster	12 27 00	+12 43	1,111	Shapley (1930b), de Vaucouluers (1953)
Shapley Supercluster	13 25 00	−31 00	14,000	Shapley (1930b): see Ch. 3
Abell SC-7	14 05 00	+27 00	44,600	Abell (1961)
Abell SC-8	14 07 00	+06 30	43,600	Abell (1961)
Abell SC-9	14 33 00	+56 30	43,600	Abell (1961)
Abell SC-10	14 39 00	+30 30	56,200	Abell (1961)
Abell SC-11	14 53 00	+22 30	43,600	Abell (1961)
Serpens-Virgo A	15 12 00	+05 30	28,333	Shane (1970), Abell (1961)
Serpens-Virgo B	15 16 18	+07 14	10,600	Shane (1970)
Corona Borealis [*]	15 23 00	+29 48	21,651	Zwicky (1957), Shane (1970)
Abell SC-14	15 32 00	+70 00	53,700	Abell (1961)
Hercules	16 02 00	+17 00	10,776	Shane (1970)
Abell SC-15	16 14 00	+29 00	10,000	Abell (1961)
Abell SC-16	23 06 00	−22 00	43,600	Abell (1961)
Abell SC-17	23 24 00	−22 30	46,700	Abell (1961)

[*] This object is also Abell SC-13 from Abell (1961).

distribution, with a preferred distance between cluster cores in the range of 200 million light-years (60 Mpc). (This value is adjusted to reflect the modern scale with a Hubble constant of 72 km s^{-1} Mpc^{-1}, whereas Abell derived it using the old scale with 180 km s^{-1} Mpc^{-1}.) Abell applied the term "second-order clustering" to what he had discovered, but today his analysis is equivalent to proving that rich clusters show a strong tendency to belong to superclusters. Unlike the qualitative statements of Shane, Abell proved his case with a statistical analysis.

George O. Abell (1927–1983)

George Abell was born, raised, educated, and worked his entire life in Southern California. He was energetic, had a congenial personality, and approached observational cosmology with open-minded enthusiasm that was infectious. After finishing his PhD at Caltech in 1957, he joined the Astronomy Department at UCLA where, with other talented colleagues, he helped build the stature of the UCLA astronomy program. Abell spent his energy on a variety of activities, which included writing popular undergraduate textbooks for introductory classes in astronomy, teaching high school science students at the Thatcher School in Ojai, and debunking popular myths in pseudo-science and astrology. After the success of his textbook, he was financially independent. Abell was a respected referee of journal papers, and immediately before his premature death at age 56, he had been selected to become editor of the *Astronomical Journal*. Abell's greatest scientific contribution was his all-sky catalogue of rich clusters. The northern sample came directly from the original plates of the National Geographic Palomar Observatory Society Sky Survey, and the southern sample came from the original plates of the UK Schmidt telescope at Siding Springs, Australia. Along with C. D. Shane and G. de Vaucouleurs, Abell was an early advocate of galaxy super-clustering. He was the first to use the galaxy luminosity function – and its natural break at L^* or M^* – to determine distances to rich clusters. Abell expressed disappointment at not having been offered a staff astronomer position at the Mt. Wilson and Mt. Palomar Observatories immediately out of graduate school, but by accepting a position at UCLA, he maximized his impact as a teacher and a mentor. Ref.: Abell (1977). Photo reproduced with permission: copyright Andrew Fraknoi.

4.3   Fritz Zwicky's Cluster Catalogue

Even though Shane and Abell were advocates of superclustering, in the 1950s not everyone agreed. The following quote written in 1957 comes from Caltech professor Fritz Zwicky:

> Restricting our analysis to those fields which do not contain any large nearby clusters of galaxies, we find that the centers of the distant clusters are distributed entirely at random. There is therefore no evidence whatsoever for any systematic clustering of clusters. . . . There exist of course accumulations of clusters of galaxies such as that in

> Pisces-Perseus or the grouping of half a dozen clusters near the cluster in Corona Borealis and its close companion. The frequency of such condensations is, however, of the order of magnitude to be expected for accidental condensations in a random field of non-interacting objects.

These words are from Zwicky's book *Morphological Astronomy* (Zwicky 1957, pp. 165–6). I had the good fortune of learning about galaxies, clusters of galaxies, and cosmology directly from George Abell at UCLA. When he spoke about Fritz Zwicky in his lectures in the spring of 1969, he showed a certain level of satisfaction in having proven Prof. Zwicky incorrect by identifying the Abell second-order clustering.

In 1969, Abell showed a similar level of satisfaction a second time during his class lectures, this time regarding Zwicky's six-volume catalogue containing both individual galaxies and clusters of galaxies. But it requires some explanation. Zwicky worked with several collaborators on the National Geographic Palomar Observatory Sky Survey images to create his widely circulated "Catalogue of Galaxies and Clusters of Galaxies." Although Abell had a clear numerically defined recipe to define his rich cluster sample, Zwicky did not. According to Abell, Zwicky adamantly claimed to have the innate ability to draw "free hand" the outer boundaries to define the Zwicky clusters, thereby encircling all reasonable cluster members. These boundaries were irregular and often complex in shape. Abell said to his UCLA class in 1969 that the Zwicky cluster boundaries are useless (perhaps he said: "carry no significance") based on the following test. In the early days, after Zwicky had begun to work on his catalogue, Abell secretly entered Zwicky's work room one night and removed from that work room the most recently finished hand-drawn catalog page. It was in Zwicky's standard catalogue format, showing all bright individual galaxies (as small circles, squares, and triangles) as well as the hand-sketched cluster boundaries. Zwicky made one map for every National Geographic Palomar Observatory Sky Survey field. Abell said nothing to anyone else, and he simply saved the original hand-drawn chart. Upon returning to work the next day, Zwicky and his assistants were forced to redraw the missing catalog page. When the final catalog was completed and published, Abell pulled from his files the page he had removed from Zwicky's workroom, and compared it to the officially published Zwicky version dated 1961. Concerning especially the cluster boundaries, Abell saw little to no resemblance between the two maps. In 1969, Abell did not show the two versions of the hand-drawn cluster boundaries, and for many years this somewhat unfinished story rested on my mind. Then in the mid-1980s, I purchased a copy of a monograph entitled *Extragalactic Astronomy* by Vorontsov-Velyaminov. Figure 7.30 in Vorontsov-Velyaminov (1987, p. 484) shows two versions of the same field from Zwicky's catalogue. In this case as well, there is little to no correspondence between the two sets of cluster boundaries, especially for the nearby Zwicky galaxy clusters. This fact is pertinent to the discussion in Chapter 6 of this book where the two Zwicky maps featured in the Verontsov-Velyaminov's book are displayed adjacent to each other on page 123.

Fritz Zwicky (1898–1974)

Fritz Zwicky received his PhD in physics in 1922 from ETH Zurich and moved to the United States from Switzerland in 1925 on a Rockefeller Fellowship. He worked at Caltech for Robert Millikan who, just two years earlier, had won the Nobel Prize in Physics. Soon thereafter, Zwicky accepted a professors position at Caltech where he remained his entire professional career. Zwicky was a close confidant of Edwin Hubble. They shared a common view of the homogeneity of the Universe and the nature of the redshift. Most of Zwicky's colleagues labeled him a maverick, insufferably arrogant, and difficult to work with. Zwicky was also a member of the staff of Mt. Wilson and Mt. Palomar Observatories. He was a founding member and research director of Aerojet Engineering where he helped develop jet engines. His major contributions to astronomy include analyzing the dynamics of the Coma cluster of galaxies and hypothesizing that dark matter is required to stabilize the cluster. In 1934, with Walter Baade, he hypothesized that supernovae represent the transformation of a normal star into a neutron star and that cosmic rays are produced in the process, ideas that all were confirmed over the next 50 years. With a group of assistants, Zwicky produced the six-volume Catalogue of Galaxies and of Clusters of Galaxies based on the National Geographic Palomar Observatory Sky Survey photographs. The individual galaxies in this catalogue, along with his galaxy magnitudes, were a perfect resource for galaxy redshift surveys in the 1970s and 1980s. Zwicky also compiled a Catalogue of Selected Compact Galaxies and Post-Eruptive Galaxies, which he published himself and distributed personally in the 1970s only to those who were not his political adversaries. This restriction continued after Zwicky's death, as he gave to his daughters a list of those who did not have permission to receive it. Today, this catalogue is available online at NASA/NED. After World War II, Zwicky was involved in humanitarian efforts to assist orphanages and to rebuild libraries around the world. Ref.: Greenstein (1974). Photo by permission: Caltech Image Archives.

Even though Zwicky may have had minimal skill in the identification of nearby galaxy clusters on the National Geographic Palomar Observatory Sky Survey photographs, his other contributions to astronomy are legendary. For finding the first evidence of dark matter in the Coma cluster, and for recognizing it as such (Zwicky 1933, 1937), he takes his place in history as an eminent astronomer. But of key interest to this book are his views regarding the galaxy distribution, and as mentioned earlier, Zwicky consistently supported a model jointly proposed by Hubble and Zwicky, in which 90% of all galaxies are members of a homogenous "field population of galaxies" and the other 10% are in occasional randomly placed clusters. As I shall describe later in the book, this model is of direct significance to the discovery of cosmic voids because one of my former redshift survey collaborators, Guido Chincarini, learned the full details of this model during discussions he had with Zwicky in Pasadena in 1972. Chincarini took this model so seriously that over many years, his aim was to measure the density of field galaxies and to determine how far into the field galaxy population he could trace the outskirts of the Coma cluster. As described in Chapter 5, this became a meaningless exercise: the Hubble–Zwicky model was ill-conceived.

4.4   Gerard and Antoinette de Vaucouleurs: Our Local Supercluster

Because of the historically significant work of Herschel and Shapley, the Local Supercluster is certainly the first object to have been identified as a supercluster. Herschel saw it as a local inhomogeneity, and Shapley saw it as significant collection of galaxies surrounding the Virgo cluster. As discussed earlier, for a short time, Shapley even mistook the Local Supercluster to be an analog of the Milky Way galaxy, as he incorrectly envisioned it in 1930. However, Gerard de Vaucouleurs made the next big step in our understanding of the galaxy distribution when he presented, first in 1953 and again in 1956, simple sketches showing external 3D views, or they might be called external 3D maps, of the Local Supercluster (de Vaucouleurs 1953, 1956). When doing so, he was trying to reconcile a report published by Rubin (1951) who suggested that the velocities of local galaxies analyzed across the sky showed evidence for systematic rotational motion. But de Vaucouleurs could find no relationship between the direction of Rubin's reported rotational motion and the geometry of the Local Supercluster. One might have expected it to be related to the central plane of the Local Supercluster.[1]

By 1953, Gerard de Vaucouleurs and his wife Antoinette were already collecting and systematically cataloging information on the Shapley–Ames sample of galaxies, the brightest 1,250 galaxies in the sky: galaxy Doppler velocities (i.e., redshifts), diameters, and precise electronically measured brightness. Once the de Vaucouleurs moved to the University of Texas in 1960, they began to make

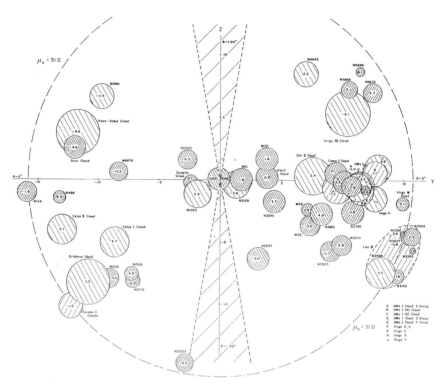

Figure 4.1a With permission: copyright 1975 University of Chicago Press

additional galaxy redshift observations using telescopes at the University of Texas McDonald Observatory. Galaxy redshift work was slow and tedious because they relied – like Slipher and Humason – on photographic plates as their means of recording the spectra. By 1964, all of their compiled information on these 1,250 galaxies was published in a catalogue called the Reference Catalogue of Bright Galaxies (RCBG). In this same era, de Vaucouleurs recognized that virtually all galaxies in his RCBG sample could be assigned membership in discrete galaxy groups or looser structures he called clouds. By 1965, de Vaucouleurs defined 15 nearby and 40 more distant groups or clouds, but all of them were part of our Local Supercluster centered on the Virgo galaxy cluster. With the Doppler velocities of approximately five members in each group, de Vaucouleurs estimated group distances based on Hubble's velocity–distance relationship. Then he created the first detailed 3D map of the Local Supercluster. His map is reproduced in Figure 4.1a and Figure 4.1b.

The maps in Figures 4.1a and 4.1b represent a major step forward in our understanding of our extragalactic neighborhood. For the first time, significant and indisputable 3D clumping was apparent in the galaxy distribution. While

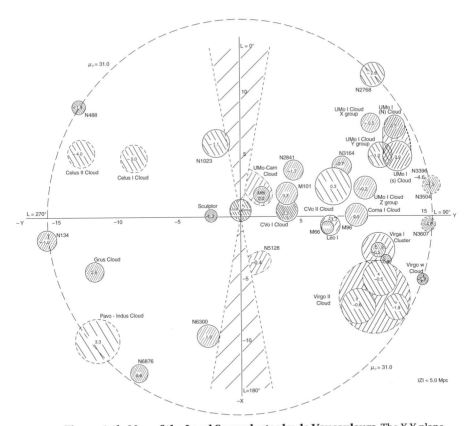

Figure 4.1b Map of the Local Supercluster by de Vaucouleurs. The X,Y plane
coincides with the flat plane of the Local Supercluster (right panel) and the Y,Z plane (left
panel) is perpendicular to it. Our Local Group of galaxies (containing the Milky Way and
the Andromeda galaxy) sit at the center the diagram (0,0,0). The core of the Local
Supercluster (the Virgo galaxy cluster) is at coordinates (−3,+12.5, −1). Circles represent
galaxy groups or clouds; the crosshatched wedges are areas obscured by our Milky Way
galaxy. The scales are given in units of Mpc. Galaxies constantly "flow" through space, so
the group positions are distorted on the order of ~30% by these flow velocities. Modern
distance calibrations show the center of the Virgo cluster is not at the 13 Mpc shown in
this diagram but at 16 Mpc. Figure credit: G. de Vaucouleurs, in "Galaxies and the
Universe," 1970, University of Chicago Press, Ch. 14, pp. 557–600. With permission:
copyright 1975 University of Chicago Press.

many astronomers, like Zwicky, continued to deny its significance, de
Vaucouleurs calculated that only 10% of the galaxies in his RCBG were not
associated with his 55 groups or clouds. With hindsight, today we can look at
this diagram and easily see that the boundaries of the Local Supercluster are
defined by cosmic voids. But this concept or terminology was not recognized in

1965. The problem is one of perspective. The supercluster systems that exist beyond (and around) the Local Supercluster were not defined. There were too few measured galaxy redshifts beyond the edge of the maps shown in Figure 4.1 to delineate additional structure.

Gerard de Vaucouleurs (1918–1995) and Antoinette de Vaucouleurs (1921–1987)

From his earliest years, G. de Vaucouleurs was interested in astronomy. As a child, he published extensive observations of planets. He received an undergraduate degree at the Sorbonne Research Laboratory in Paris 1939, but life became complicated with the onset of World War II. When the French government collapsed, he retreated to a private observatory in the south of France where he continued to work and learn astronomical techniques. He returned to Paris 1943–1949 to work on his dissertation (on molecular Rayleigh scattering of light) at the Laboratory of Physics Research and the Institute d'Astrophysique Boulevard Arago. During his time in Paris, he met and married Antoinette. She also studied at the Sorbonne. In 1949, they moved to London where Gerard worked at the BBC. In 1951, he left for the Australian National University and Mt. Stromlo Observatory as a research fellow. In Australia, his extragalactic work on the Local Supercluster began, and he also worked on the Magellanic Clouds. By early 1957, he accepted a position in Flagstaff AZ at Lowell Observatory (based on his early work on Mars). When that appointment stumbled, he moved briefly to Harvard, and soon afterward (in 1960), Gerard and Antoinette moved to the University of Texas where they spent the rest of their careers. Together, they are best known for completing the massive *Reference Catalogue of Bright Galaxies* (editions in 1964, 1976, and 1991; the last two with collaborator H. G. Corwin, Jr.). These catalogues contain detailed information on 2,599 galaxies, 4,364 galaxies, and 23,024 objects, respectively. The first catalogue was the basis for the 3D investigation of the Local Supercluster. In the late 1950s, Antoinette de Vaucouleurs was the first to recognize that active galaxy nuclei fluctuate in brightness. Gerard de Vaucouleurs was well known for his competition with Allan Sandage on the distance scale. De Vaucouleurs advocated a large value for $H_0 = 100$ km s^{-1} Mpc^{-1} and Sandage a small value $H_0 = 50$ km s^{-1} Mpc^{-1}. Sources: Burbidge (2002); de Vaucouleurs (1988, 1991). Photograph by permission: copyright and photo-credit McDonald Observatory/UT-Austin.

De Vaucouleurs considered the large-scale structure beyond the Local Supercluster, but rather than postulate that the Universe is filled with superclusters and cosmic voids, he went in another direction and began to speculate that the level of irregularity in the galaxy distribution might continue to increase as we probe deeper into the Universe (de Vaucouleurs 1970, 1971). In other words, the size of the structures and the size of the empty regions might get bigger and bigger as the sample volume increases. He was aware that Swedish astronomer C. V. L. Charlier (1862–1932) had suggested in 1908, an alternate cosmological model that was based not on a homogeneous matter distribution but on one that showed a continuous clustering hierarchy: as the scale increases, the level of irregularity increases (Charlier 1908). In the same time frame that de Vaucouleurs published his views on hierarchical structure, James Wertz was awarded a PhD degree in astronomy from the University of Texas. His thesis was entitled *Newtonian Hierarchical Cosmology* (Wertz 1970, 1971). This development looked promising to de Vaucouleurs and Wertz, until Allan Sandage and his collaborators, in 1972, used Wertz's mathematical predictions and Sandage's observations to decisively invalidate the updated version of the model with a clustering hierarchy (Sandage, Tammann, & Hardy 1972). After being rebuked by Sandage et al. in 1972, de Vaucouleurs seemed to halt further speculation regarding unconventional cosmological models. Wertz appears to have done the same.

4.5   Holdouts for Homogeneity

The general nature of the large-scale galaxy distribution would not be revealed until the end of the 1970s when significantly deeper galaxy redshift surveys displayed clear evidence of 3D structures outlining the network of superclusters and cosmic voids that lie beyond our Local Supercluster. But in the meantime, there was still a window of opportunity for those who maintained, like Zwicky did, that the galaxy and the galaxy cluster distributions might be homogeneous. I will end this chapter with two examples.

In this period, Princeton University Professor P. J. E. Peebles seemed to work the hardest to maintain the traditional view that the local galaxy distribution is homogeneous. Like most scientists in that day, he never insisted on this outcome but instead suggested it firmly. For the moment, one example will suffice. In 1969, Peebles wrote a paper with his student Jer Yu, in which they considered whether the clumped "second order clusters" reported by Abell in 1958 are truly clumped or whether these are simply chance statistical fluctuations (Yu & Peebles 1969). During their analysis, they began to suspect errors in Abell's methods of cluster discovery because they saw a change in the number of clusters with galactic latitude. Second, they noticed that most of the rich cluster clumping occurred in a group of more distant objects, in what Abell called "distance class 5" rich clusters. In all other groups of data (one group more distant than class 5 and the others nearby), the clumping was much weaker, so they dropped the "distance class 5" objects from their analysis, and by doing so, the superclustering tendency was sharply reduced. Abell was displeased with their analysis (Abell 1977). Even de Vaucouleurs got involved when he said about this situation "it is difficult to understand how diametrically opposite conclusions could be reached from the same data – Abell's catalog of rich clusters – by Yu and Peebles (1969)" (de Vaucouleurs 1971). Peebles continued to investigate Abell's second-order clustering, and eventually reversed his stance (Hauser & Peebles 1973), and Abell (1977) claimed that he was eventually "vindicated." This example characterizes the status quo in the late 1960s and early 1970s. There was considerable uncertainty. The ultimate answer to this problem would come from those of us who later greatly expanded the redshift surveys into the deeper Universe.

The second example is a paper published in January 1976, by the late Princeton Physics professor John Bahcall (1934–2005) and his collaborator Paul Joss (b. 1945), who was, at the time, a young physics professor at MIT. Bahcall and Joss (1976) suggested that the Local Supercluster may not be a real physical system in the sense that de Vaucouleurs had proposed, but instead that it could be explained as a statistical fluke or irregularity in a relatively uniform galaxy distribution. This is exactly the manner used by Zwicky to discuss the insignificance of other superclusters. Bahcall and Joss (1976) specified certain conditions or assumptions within which their idea would work. These included confining galaxies to clouds of galaxies resembling those de Vaucouleurs had observed, that these clouds be randomly situated around the Virgo cluster, and that the radius of the Virgo cluster itself be at least 15°. Bahcall and Joss talked at length in their paper about the long-range gravitational effects of the Local Supercluster and how these effects might induce galaxy "flows," thus distorting our perception. Because these flows were actually detected years after their paper was published, the reality of the Local Supercluster could not be challenged on this basis today. But their paper is still a curious fossil of history. Bahcall and Joss were siding in 1976 with Zwicky's earlier views by suggesting that the galaxy distribution in the local Universe might be a somewhat clumpy but random 3D distribution.

These two examples demonstrate the uncertainties that existed in the cosmology community into the mid-1970s. The Bahcall and Joss paper appeared just two years before Gregory and I published our Coma/A1367 Supercluster paper in 1978. There were many astronomers who would not accept the level of inhomogeneity that de Vaucouleurs was detecting in the

Local Supercluster. This attitude persisted, despite the fact that lists of candidate superclusters (as given in Table 4.1) had been published. Very few astronomers and cosmologists were prepared for the even more extreme inhomogeneities that our future 3D redshift maps would soon show. As most readers can anticipate, our results would also generate doubt and would be ignored by those who shared the views of well-established scientists like Zwicky, Peebles, and Bahcall. Ironically, I encounter astronomers today who look back at older research results and claim that cosmic voids were apparent long before 1978. But by doing so, they ignore the context. They ignore the persuasive attitudes of leaders in astronomy and cosmology like John Bahcall and P. J. E. Peebles who were hesitant in the 1970s to step away from the old models of a very nearly homogeneous galaxy distribution.

5

The Early Redshift Surveys from Arizona Observatories

In the late 1960s, the Astronomy Department at the University of Arizona began to move into the upper echelon of astronomy programs in the United States. Many factors were contributing to the explosive growth of astronomy in Tucson. Located just across the street from the Astronomy Department on North Cherry Avenue was the headquarters of Kitt Peak National Observatory (KPNO). The University of Arizona's Lunar and Planetary Laboratory and the Optical Sciences Center were nearby. In 1966, Bart J. Bok (1906–1983) had been appointed Head of the Astronomy Department and Director of Steward Observatory (Graham, Wade, & Price 1994). New telescopes peppered the central ridge of Kitt Peak, just 50 miles to the southwest of the town. The Steward Observatory's 36-inch telescope, originally located in a dome on Cherry Avenue, had been moved to Kitt Peak and placed adjacent to the site set aside for the Steward Observatory's new 90-inch telescope. In June 1969, the 90-inch telescope (it is now called the Bart J. Bok 2.3-m Telescope) was dedicated and put into service. Even more significant was the impending completion of the KPNO 150-inch telescope slated for 1973. It would provide the means to compete with the Mt. Palomar 200-inch telescope in cosmological studies with a wide-field prime focus camera, making it superior in that regard to the 200-inch.

Stephen Gregory and I began our graduate studies at the University of Arizona in the 1969/1970 academic year. My undergraduate advisor at UCLA told me that there was no better graduate school for a student who aimed to become an observational astronomer. I was admitted to the Arizona astronomy program and was awarded a three-year scholarship (funded by the National Defense Education Act), which gave me great flexibility. Gregory was admitted and supported financially through research assistantships. I eventually realized that to work side-by-side with designated professors, as happened with Gregory, may have provided better training than to be free in the way I was. The courses taught by the professors were acceptable, but I learned more by direct means: using my time to observe objects with the telescopes on Kitt Peak and to read research papers, tracing important ideas back in time, starting with references that were listed in the newest publications.

When Bart Bok arrived at the University of Arizona in 1966, one of his priorities was to coordinate technology development in preparation for the completion of the Steward 90-inch telescope. Bok had just come from Mt. Stromlo Observatory in Australia where he had similar responsibilities. The previous Steward Observatory director, Aden Meinel, set a solid foundation. The observatory was preparing for the installation of an image-intensified camera for spectroscopy incorporating an image tube detector pioneered by H. Kent Ford Jr. at the Carnegie DTM (described in Chapter 1). University of Arizona graduate student Richard Cromwell was awarded his PhD degree in early 1969 for building the image-intensified camera system for Steward Observatory, and Cromwell continued as a member of the observatory staff to place this system into smooth operation. The first scientific results from the new spectrograph came in late December 1969 and early January 1970 at the 90-inch telescope (Cromwell & Weymann 1970). Arizona professor William G. Tifft had a separate instrument development program (funded by NASA) to build another advanced camera system called an "Image Dissector Scanner" (Tifft 1972a). This instrument was also used at the 90-inch telescope, but its performance was no match for the image tube systems that became the first "work horse" instrument at the 90-inch telescope.

5.1   Preliminary Research in Graduate School

In the spring after arriving, I arranged to use the Steward 36-inch telescope on Kitt Peak for 15 to 20 nights to monitor the brightness of galaxy nuclei with a photoelectric photometer. One of my targets was the Seyfert galaxy NGC 4151. This project was suggested by and done in conjunction with Arizona Professor Andrej Pacholzyk. It was my first experience at a major observatory. Meanwhile, Steve Gregory began working with Professor Ray Weymann on galaxy spectra from Seyfert galaxy nuclei obtained at the 90-inch telescope. This was Gregory's first serious step into research. He would later switch to work with Professor William Tifft for his PhD thesis to collect redshifts of galaxies in the Coma cluster. While doing so, Tifft and Gregory formed a close collaboration.

For a short time after I arrived, I considered working with the venerable Bart Bok on projects involving our Milky Way galaxy. Bok had a diverse research portfolio, and in the spring of 1970 he had organized a small meeting at Steward Observatory where presentations were made by his group members. I participated in the meeting; some of those who attended are shown in Figure 5.1.

Meanwhile, I became fascinated with rich Abell galaxy clusters, made my own separate catalogue of the nearest clusters, and graphically mapped their distribution on the sky. In my first year of graduate school, I also read papers by Refsdal (1964) on gravitational lensing and decided to test the idea that dark matter in galaxy clusters might be identified by looking for what are called "Einstein rings." If dark matter in Abell clusters consisted of massive discrete objects, Einstein rings (the distorted images of background galaxies) might appear in the direction of the cluster. I did a preliminary search for this effect on plate copies of the National Geographic Palomar Observatory Sky Survey located in the basement of the KPNO headquarters across North Cherry Avenue, and the results came out negative in the following sense. I did see ring-like objects associated with the clusters, but the rings were not distributed like the galaxies and the dark matter: most rings were in the cluster outskirts. So I abandoned that idea as a thesis project and later published my ring galaxy identifications under a different guise. For my PhD thesis work, I eventually

Figure 5.1 Bok Symposium 1970. Left to right: Carolyn Cordwell McCarthy, Maxine Howlet, Bart Bok, Robert Elliott, Priscilla Bok, Raymond White, Nannielou Dieter, Arthur Hoag, Beverly Lynds, Robert Hayward, William Fogarty, Jack Sulentic, Ray Weymann, and Laird A. Thompson. With permission: copyright Steward Observatory, University of Arizona.

settled on a project to investigate the rotational properties of individual galaxies in rich Abell clusters by looking for systematic trends in the shape and in the alignment of the rotation axes of the member galaxies. Today, this popular field of research goes under the name of "galaxy intrinsic alignment." Apart from the observational aspects of my thesis, one requirement was my need to become familiar with current theoretical models of galaxy formation. Several of these models came from Soviet astrophysics. Among them were the models of Zeldovich. Perhaps the most interesting and, to me, influential model was described in a paper written by Sunyaev and Zeldovich (1972). By early 1972, I had also selected William Tifft as my thesis advisor, and I had defined the path for completing my PhD degree.

During my second year of graduate school, several serious complications arose. First, the American Astronomical Society (AAS) – the professional organization for career astronomers in the United States – projected in 1971 that there were too many graduate students being trained relative to the number of jobs that would be available in the upcoming two decades. The AAS suggested that it was the duty of individual astronomy departments to reduce the number of existing students in PhD granting programs. This raised the level of student-to-student competition. Second, the Vietnam War was at its height, and while college students were demonstrating against the war on campuses across the country, the US Congress ordered a lottery to designate who would be drafted to fight in Vietnam. My lottery number came out to be six, but I had arranged a slot for myself in the US Army National Guard. The day after the lottery, I joined the National Guard to be trained as a military policeman. This delayed my education a total of approximately one year, from my cumulative military service, in the period 1970–1976. One fellow Arizona graduate student was a veteran; two were in the Reserve Officer Training Corps; another, a conscientious objector; but Gregory and most other students did not have to serve. Finally, in 1971 or 1972, our PhD thesis advisor William Tifft began to discuss his new alternate way to interpret galaxy redshifts. Based on his observations of galaxies in the Coma cluster of galaxies (see Figure 5.2), Tifft claimed that galaxy redshifts are "quantized" (e.g., Tifft 1972b, 1976). By this he meant that galaxy redshifts did not fall along a continuum of values but came in bunches or groups. When he first identified this apparent phenomenon in the Coma cluster, he called the groupings "redshift bands." If this apparent effect had been proven to be true, it would have been revolutionary. If the concept were wrong, Tifft would lose considerable credibility. Jobs were going to be scarce in the next decade, and successful students would need strong support from a strong PhD advisor. I saw quantized redshifts as a serious obstacle. Gregory accepted the concept as a real possibility and agreed to be a coauthor with Tifft on several research papers that discussed it. I kept as far away from it as I could. On only one point did I agree with Tifft: he had a right to interpret and publish his observations of galaxies as he saw fit, no matter how radical his interpretation might be. William Tifft has never stepped back from the idea that galaxy redshifts are quantized even though nearly 50 years have now passed since he first proposed it. Tifft's redshift bands have neither been proven nor disproven, but most astronomers feel comfortable to ignore them and label the banding effect visible in Figure 5.2 a fluke.

Tifft's other research thrust in astronomy was a conventional investigation of the rich Coma cluster of galaxies. He took on the task of collecting and measuring hundreds of galaxy redshifts for this study. One night after another, the intensified spectrograph camera on the 90-inch telescope recorded many galaxy spectra, and Tifft enlisted the help of Gregory to measure these spectra and thereby compile a unique data set for the Coma cluster. The project was in some sense open-ended because the Coma cluster is composed of several thousands of galaxies. To keep the project at a reasonable size, Tifft and Gregory's goal was to extend the survey to an angular distance of 2.79° from the cluster

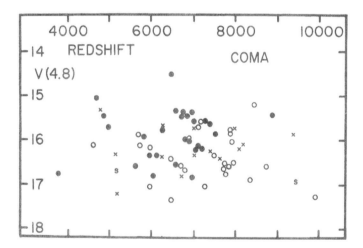

Figure 5.2 Tifft redshift-magnitude diagram. When Tifft published this diagram in 1972, he became an advocate of unconventional physics. The diagram plots the observed galaxy redshift on the horizontal axis and the brightness in the central core of the galaxy (the nuclear magnitude) on the vertical axis. The primary basis for his early suggestion that galaxy redshifts are quantized is the appearance of the diagonal bands visible in this diagram. Tifft spent more than 30 years pursuing non-cosmological redshift phenomena and has published more than 25 research papers discussing it and related topics in the *Astrophysical Journal* and elsewhere. By permission of the A.A.S.: W. Tifft (1972), *Astrophys. J.*, 175, 613–35.

center. Galaxies were selected from the Zwicky catalogue to insure a brightness-limited survey. The analysis and interpretation of this inner sample became Gregory's PhD thesis (Gregory 1975).

William G. Tifft (b. 1932)

William Tifft, a top-10 National Science Talent Search winner as a high school senior, entered Harvard College receiving his AB degree Magna Cum Laude in 1954. His 1958 PhD thesis at Caltech involved detailed photometry of galaxies. A two-year NSF post-doctoral fellowship at Mount Stromlo Observatory in Australia was followed in 1960 by a Research Associate position at Vanderbilt University with Professor Carl Seyfert. However, Professor Seyfert died in an auto accident one hour before Tifft's initial arrival. Tifft's entire career was built to understand precision studies of real data to expand and test current theories. He accepted an Astronomer position at Lowell Observatory in 1961 where he extended precision photometry of both radial and nuclear properties of galaxies. In 1964, he joined the University of Arizona as an Associate Professor to develop and operate a Space Astronomy Laboratory. For a short time in 1965, he was one of sixteen initial NASA Science Astronaut applicants. In 1973, as a Full Professor at Steward Observatory, he turned full time to research, teaching, and developing a major study of the redshift to fundamentally test classical cosmology. An initial study in 1970 indicated that redshifts seemed to contain an intrinsic aspect, and his work diverged from classical cosmology. In a basic study of the Coma cluster, Tifft (1972) revealed a redshift-magnitude non-dynamical banded pattern confirming his finding that the redshift was indeed a quantized unit. Follow-up with double galaxies led to global studies, external confirmation by Guthrie and Napier (1991), and many details. He retired as Professor Emeritus in 2002 with his Cardiff presentation of Quantum Temporal Cosmology (QTC) described in "Redshift Key to Cosmology" (Tifft 2014). Tifft was recognized by Who's Who in the 2017–2018 top Achievement Award. QTC is built upon 3D time and the apparent understanding of space-time structure. Photo reproduced with permission: copyright William Tifft.

Gregory was the first in our class to finish his graduate degree. Just before he left Tucson for a new professor's appointment in upstate New York, he suggested to me that he and I together should submit telescope time requests for the KPNO telescopes to study the morphology of galaxies in rich Abell clusters and to measure redshifts of the member galaxies. This project would take good advantage of our combined skills. The galaxy morphology study would require the new 150-inch telescope and since it was not yet available (and would not be available for another year), these became plans for the future. I agreed to the collaboration and was pleased that Gregory had suggested it. Up until that point we had been competitors. It made sense that we join forces.

During the next year, the KPNO staff thoroughly tested the 150-inch telescope before it was opened to outside users. While I was completing my PhD research, at times I would casually visit with the astronomers on the other side of North Cherry Avenue. I began to see gorgeous research photographs from the prime focus camera. Gregory and I eventually submitted our first joint observing proposal, and it was our good luck that, despite the very competitive nature of the telescope time requests, we were notified in October 1974 that we had been granted three nights on the 150-inch telescope for early May 1975 to take wide-angle prime focus photographic plates of rich galaxy clusters. The second half of our request – for telescope time to measure galaxy redshifts – failed, but we received positive feedback suggesting that we could resubmit it at a later time.

In those days, time passed very quickly. After finishing my PhD degree in late 1974, I accepted a research position to work with KPNO staff astronomer Stephen Strom. I published the main results from my PhD thesis (Thompson 1976) and also a paper on "Ring Galaxies in Rich Clusters" (Thompson 1977). It made no mention of my failure to detect dark matter. Gregory was in New York teaching at a small state university, and he returned to Tucson in the summers to continue his work with Tifft on extending the Coma cluster redshift survey. At first, Tifft and Gregory (1976) extended their survey to a radius of 3° and eventually to a radius of 6°.

Herbert J. Rood (1937–2005) and Guido Chincarini (b. 1938) began to collect galaxy redshifts in the Coma cluster at just about the same time that Tifft and Gregory began their Coma cluster survey work. While Tifft and Gregory primarily used the Steward Observatory 90-inch telescope, Chincarini and Rood used the KPNO 84-inch telescope. As mentioned in Chapter 2, both telescopes were equipped with similar "state-of-the-art" spectrographs with image-intensified cameras to amplify the galaxy spectra. As users of the KPNO telescopes, Chincarini and Rood relied on the observatory's telescope time allocation committee (TAC) to grant them time in the standard competitive process. During this period, Rood was a professor at Michigan State University and Chincarini was a professor at the University of Oklahoma. Five years before their collaboration began, Rood developed a strong interest in rich galaxy clusters that started with his PhD thesis in 1965. Chincarini, who was more of an equipment-oriented astronomer, joined Rood in continuing his work on galaxy clusters.

Guido Chincarini (b. 1938)

Guido Chincarini began his research career at Asiago Astrophysical Observatory, University of Padua, Italy, where he received his PhD in 1960. From 1964 to 1967, he worked as a postdoctoral appointee at the University of California Lick Observatory, where he first assisted and then worked side by side with Merle Walker in using the Lallemand camera at the coudé focus (behind the 20-inch Schmidt camera) of the Lick 120-inch telescope. The Lallemand electronographic camera was the first electronically enhanced image tube camera placed in operation at astronomical observatories. From 1969 to 1972, Chincarini was a Visiting Scientist in the Thornton Page group at the NASA Manned Spacecraft Center (now the Johnson Space Center) in Houston Texas, and in 1972, he moved to McDonald Observatory, first at Ft. Davis, Texas, and later in Austin, Texas. In fall of 1976, Chincarini joined the Physics and Astronomy Department at the University of Oklahoma in Norman, Oklahoma, where he remained until joining the European Southern Observatory in 1983. In 1985, he was offered the Chair of Astronomy and Cosmology at the University of Milan and the Director of the Astronomical Observatory of Brera. Guido Chincarini is of medium height and relatively slight build. When he gets excited, the pitch of his voice goes up one or two octaves.

Herbert J. Rood (1937–2005)

Herb Rood received his PhD degree from the University of Michigan in 1965 with a thesis entitled "The Dynamics of the Coma Cluster of Galaxies." He was first a professor at Wesleyan University (Middletown, Connecticut) through 1972, and then at Michigan State University through 1979. Starting in January 1980, Rood was associated with the Institute for Advanced Study at Princeton University, sometimes as a visitor and other times as a member. This was convenient for him as it was close to his parent's home located near Trenton, New Jersey. Rood had cerebral palsy, which he largely overcame in order to dedicate his life to explore the detailed properties of the Coma cluster and other rich galaxy clusters. Rood established a world-class reputation for his work on Coma. In the early 1970s, I had the opportunity to select one outside member to sit on my PhD thesis exam committee, and I chose Herb Rood. We corresponded by US mail about this matter, but one month before the exam date, I received a hand-written personal letter from Herb Rood's father saying that his son had suffered a fall and had hurt his back: this medical circumstance would prevent his son from traveling to my exam in Tucson. A number of years later, I would learn that, in September 1974, Herb Rood sustained injuries when he tried to take his own life. Fortunately, the 1974 incident was not fatal, and Rood continued to contribute to our understanding of rich clusters of galaxies for another 30 years. Herb Rood did not have an easy life.

Chincarini and Rood's 11-year collaboration ended in 1981, and on the last three papers they published together, I worked with them as a coauthor. The three of us knew each other very well. I traveled twice to Norman, Oklahoma, to visit Guido Chincarini, and he and Herb Rood visited me once when I later moved to the University of Nebraska, Lincoln. I left Nebraska by mid-1979 and moved to the Institute for Astronomy at the University of Hawaii. In 1981, Herb flew to Honolulu. At that time, Rood, Chincarini, and I were working together on what we called the "supercluster bridge," a paper I discuss more fully later in this chapter. When we collaborated, Herb Rood was always a fair and even-tempered gentleman. To me, Guido Chincarini always came across as a competitor.

Tifft had an active research group in the 1970s and managed at least five graduate students, four technical employees, and one postdoctoral researcher. This was a time when social norms were rapidly changing, and I have always wondered whether Tifft's attraction to his radical redshift band theory was somehow linked to the social atmosphere of the early 1970s. In this era, there were heated debates between adherents of conventional cosmology and those who promoted non-cosmological redshift concepts. Two prominent non-cosmological proponents at other research organizations included Halton Arp and Geoffrey Burbidge. After 1972, Tifft was the most prominent non-cosmological redshift scientist in Arizona. On another front, Tifft showed his own personal peculiarities by wearing brightly colored sports jackets, dress pants, and socks: chartreuse, orange, lime green, and many times the colors were not well coordinated. Within his research group, the graduate students sometimes called him "Rainbow Bill."

In the mid-1970s, Tifft's postdoctoral researcher was Massimo Tarenghi (b. 1945), who had come to Arizona from Milan, Italy. Tarenghi had an office in Steward Observatory just down the hall from Tifft's office, and Tarenghi seemed to spend a majority of his time using a hand-cranked precision measuring machine to analyze the photographic plates of galaxy spectra that came from the image-intensified camera system. These measurements were among the final steps required to determine a galaxy redshift. Tifft and Tarenghi published papers together that were primarily on the redshifts of galaxies that were sources of radio wavelength radiation. Tarenghi also began to work on a sample of approximately 100 spectra of galaxies located in the Hercules supercluster.

Tarenghi and I became good friends and remained so after I received my PhD and moved across North Cherry Avenue to work at KPNO. Gregory left Tucson one month before Tarenghi arrived. When Gregory would return to Tucson for the summer, he had to notice that Tarenghi had taken what used to be his place in

Tifft's office sitting in front of the small measuring engine where galaxy spectra were measured by hand. Gregory had little interaction with Tarenghi. This is a significant point given the way the early redshift survey collaborations evolved.

5.2   Chance Encounter on Kitt Peak

A rare event happened in early May 1975, when quite by chance the three largest telescopes on Kitt Peak were scheduled to simultaneously observe the Coma cluster of galaxies. Gregory and I were using the prime focus camera on the KPNO 150-inch telescope. Tifft was using the Steward 90-inch telescope, and Chincarini and Rood were using the KPNO 84-inch telescope. Tifft, Chincarini, and Rood were all collecting galaxy spectra. It was on our last night at the 150-inch telescope (May 7, 1975) when all five of us showed up for dinner at the observatory cafeteria before our long observing night began. We met in the dining room where there were modest-sized rectangular tables. Tifft sat at the head, Gregory and I were on one side of the table, and Chincarini and Rood were on the other. Gregory and I were in our late twenties; Chincarini and Rood were in their late thirties. This was an interesting time in the development of our understanding of the large-scale structure. Gregory and I may have had the best picture as to what was going on in terms of the galaxy distribution, but the full picture was not resolved. The Gregory and Thompson view can be read directly from the telescope time request we would soon submit to observe with the KPNO 84-inch telescope to measure redshifts in the Coma/A1367 supercluster field (reproduced in Appendix A). Rood (1988b) and Chincarini (2013) both have described this meeting themselves.

In May 1975, Tifft was 43 years old. He carried himself well – somewhat like a bank president – and presented his views in a confident manner. Tifft liked to talk even in a philosophical manner about his work, but in those days, I suspect he enjoyed shocking others with his unconventional redshift theory. By mid-1975, he had already been working on the quantized redshift concepts for three to four years and had published three papers discussing this phenomenon in the *Astrophysical Journal*. Tifft and Gregory were preparing their excellent (conventional) paper on the 6° Coma cluster redshift sample. It would be submitted to the *Astrophysical Journal* in final form, approximately three months after our dinner meeting on Kitt Peak. These new results on Coma were of direct interest to Chincarini and Rood who, until that time, had put little thought into the large-scale distribution of galaxies. Instead, their focus was studying individual clusters of galaxies. At dinner that late afternoon, Tifft made a big deal of the fact that the Coma cluster foreground appears to be largely empty and on the further result that the small number of galaxies seen in the foreground are clumped in the redshift coordinate. At one point, he added wryly that they fell in discrete groups or bands. He used the latter word in reference to his non-cosmological redshift work. In the published paper by Tifft and Gregory, the foreground galaxies are discussed, but there is no direct reference to non-cosmological effects. The fact that the foreground to the Coma cluster is quite empty is discussed but is not emphasized. The most prescient conventional comment in Tifft and Gregory's published paper is in a figure caption where they say that "the greater part of the foreground is completely devoid of galaxies." The graph from Tifft and Gregory is shown in Figure 5.3.

At dinner, Chincarini and Rood described the continuation of their redshift survey work in the outskirts of the Coma cluster. They aimed to find an end to the Coma cluster: where does it smoothly merge with galaxies in the surrounding space? We learned nothing new from Chincarini and Rood. I recall walking down the sidewalk away from the dining room after our informal meeting was over, and once Gregory and I were out of earshot range, I said to Gregory, "I hope

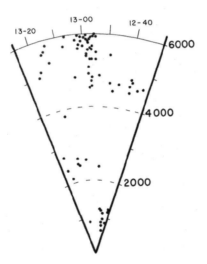

Figure 5.3 Tifft and Gregory Coma cluster redshift survey. This figure, reproduced from Tifft and Gregory (1976), is the first modern use of a wedge or cone diagram to display the 3D distribution of galaxies. Although it provides evidence that the galaxy distribution is highly non-uniform, the Tifft and Gregory survey is somewhat too narrow to reveal the structure of voids. Although the Tifft and Gregory survey extends to a radius of 6° from the cluster center (a total sweep of 12° on the sky), the opening angle in the graph is exaggerated in the plot by 3.3 times to 40°. The idea of using a cone diagram to display the data arose from a conversation Gregory had with a former colleague at the State University of New York at Oswego, Tom Edwards, who was not an astronomer. By permission of the A.A.S.: W. Tifft & S. Gregory (1976). *Astrophys. J.*, 205, pp. 696–708.

we did not reveal too much of our 3D Coma/A1367 supercluster mapping concept." We had the idea that adjacent rich galaxy clusters might be connected to each other and that we might find evidence for these connections in the 3D galaxy distribution. In fact, judging from the papers all five of us eventually published, the outcome of the meeting was reasonable and satisfactory. Chincarini and Rood had, indeed, picked up several key ideas about how to use their Coma cluster redshift data in discussions of the general galaxy distribution, but when they published their results later that year in *Nature* and in 1976 in the *Astrophysical Journal*, they did not have the foresight to recognize, let alone discuss, the hidden concept that they could use their redshift data to map the 3D structure. Looking back at their two publications today one can see evidence for the structure in the graphs of their basic data, but they did not recognize its significance in a 3D interpretation. They remained true to their defined aim: to find the edge of the Coma cluster by detecting the radial distance where the cluster merges into the smooth distribution of field galaxies, and then they ascribed the gaps in their redshift plots to "segregation in redshift" with its vague non-cosmological redshift implications.

Gregory and I successfully finished our work at the 150-inch telescope that night, and the next day we rode in the observatory shuttle through the Arizona desert, 50 miles back to Tucson. Gregory headed to the airport to return to upstate New York, and I began the initial work on the beautiful new wide-field photographic plates of the Coma cluster, the Hercules supercluster and the cluster pair A2197 and A2199. Chincarini and Rood remained on Kitt Peak for another few days to complete their nighttime observing.

5.3   Hercules Supercluster Collaboration

Over a fairly long timescale, Tifft and Tarenghi had been collecting galaxy spectra in the Hercules region, and Chincarini and Rood were doing the same. Gregory and I had just finished taking prime focus images of rich galaxy clusters and we included the Hercules region in our sample. An influential astronomer at Steward Observatory, Peter Strittmatter, recognized this overlap of interest and suggested to Tifft and Tarenghi that a group study of the Hercules supercluster would be advantageous for everyone. So Tarenghi took charge by traveling up to the observatory on Kitt Peak while Chincarini and Rood were still observing to suggest a collaboration with them. With their cooperation guaranteed, Tarenghi next talked with me. Immediately, he organized a meeting aiming to cement a collaboration. On the evening of May 10, 1975, after Chincarini and Rood had returned to Tucson and were ready to return home, Tarenghi invited me, Chincarini, and Rood to dinner at Caruso's Italian Restaurant on North 4th

Avenue in Tucson in order to ensure that all five of us agreed to work together on the Hercules supercluster: Tarenghi, Tifft, Chincarini, Rood, and me. Tifft could not attend the dinner, but he had already agreed with Tarenghi that everyone join together. Our dinner went well and the collaboration was agreed upon by all. I would provide new information from the original KPNO 150-inch prime focus photographic plates: galaxy morphology, galaxy orientation, and other static properties; Tarenghi and Tifft would eventually provide nearly 100 new redshifts; Chincarini and Rood would provide 47 more; and there were 45 other galaxy redshifts that were already published and available in de Vaucouleurs' "Reference Catalogue of Bright Galaxies."

Massimo Tarenghi (b. 1945)

Massimo Tarenghi graduated with a PhD degree in Physics from the University of Milan in 1970. After postdoctoral appointments in Milan and Pavia, he became a European Space Research Organization Fellow at Steward Observatory in Tucson from 1973 to 1975 where he undertook many projects including infrared wavelength observations of galaxies, redshift observations of radio galaxies, and most importantly, headed a collaborative study of the Hercules supercluster region. Two years after returning to Europe in 1975, he joined the science group at the European Southern Observatory (ESO) where he worked to automate the instrumentation on the ESO 3.6-m telescope's prime focus camera. When a 2.2-m telescope was redirected to ESO in the early 1980s, Tarenghi was appointed the Project Manager, and he directed its installation including making the telescope operate remotely over telephone lines. In 1983, he was asked to be Project Manager for the 3.58-m New Technology Telescope where thin primary mirror technology was pioneered by Raymond Wilson. In 1991, Tarenghi was appointed Project Manager for the ESO Very Large Telescope and immediately afterward became Director of Paranal Observatory where he stayed until 2002. Then, in 2003, he became Director of the new Atacama Large Millimeter Array (ALMA). For his exceptionally successful career at ESO, Tarenghi has won many awards. Throughout his career, he maintained an interest in the Hercules supercluster and in radio galaxies associated with rich clusters of galaxies. Refs.: Madsen (2013) and Catapano (2015). Photo reproduced by permission: copyright Massimo Tarenghi.

Our agreement to work together on the Hercules supercluster was a good plan, but it would take longer to complete the entire project than anyone anticipated on that evening in May 1975. Before the Hercules study made significant progress, Gregory and I would be granted KPNO telescope time on the 84-inch telescope to collect galaxy redshifts within an area 24° by 15° encompassing the Coma/A1367 supercluster, and we would complete our observations and even submit our paper for publication. The delays in the Hercules supercluster project were caused by our group leader, Massimo Tarenghi, who would, in the meantime, resign his post-doctoral position in Arizona to accept a position of increasing responsibility at the European Southern Observatory headquarters in Munich, Germany. Tarenghi moved to his new job before he had finished measuring all of the Tifft and Tarenghi galaxy spectra from the Hercules supercluster.

Starting in May 1975, when the five of us had dinner on Kitt Peak, Chincarini and Rood had to be worried about receiving future telescope time at KPNO. The telescope time allocation committee (TAC) kept an eye on those astronomers who had been given telescope time in the past. The TAC had been generous to Chincarini and Rood for ~5 years, and yet the papers they had published up until that point had limited content. By May 1975, Gregory and I had already established considerable forward momentum. When we were granted time to obtain redshifts in the Coma/A1367 supercluster, Chincarini and Rood are likely to have submitted their own proposal and to have come up empty handed. Gregory and I were not aware of this potential problem in the 1970s, but in the personal account written 38 years later, Chincarini (2013) reveals his frustration with the TAC.

5.4 Coma to A1367: Discovering Cosmic Voids

Observing projects directed at single objects like the Coma cluster are annual affairs because any specific object slowly drifts, night by night, into the ideal position for observing. The Coma cluster is best observed in early spring, and the rich cluster A1367 is in its best position a few weeks earlier. By late April of the next year (1976), Gregory and I were back on Kitt Peak at the 84-inch telescope to collect galaxy spectra for our Coma/A1367 redshift survey. Because Tifft and Gregory found a clumpy galaxy distribution in the Coma cluster fore-ground, we were excited to see how galaxies were distributed in the 24° long swath of sky that contained the two rich galaxy clusters: we would be sampling one of the more interesting regions of the local Universe. The volume in question lies – in terms of its distance – between the Local Supercluster (the Virgo cluster sits at a distance of 50 million light-years (16 Mpc)) and the Coma/A1367 complex at a distance of 300 million light-years (95 Mpc). The questions in our minds were: Are the structures defined by the galaxies in this volume of space somehow distinct or different? And exactly what lies in the region of space between the two rich clusters, Coma and A1367? Is there a "bridge" of galaxies connecting the two clusters?

Anyone who studies the distribution of galaxies over wide angles in the sky recognizes the profound nature of these investigations. Galaxies move dynamically under the force of gravity, but they move small distances in a relative sense. Galaxies close to each other on the sky also tend to move in unison (i.e., they flow) in response to the over-dense and the under-dense regions around them. Given the age of the Universe at 13.8 billion years, galaxies at the distances of Coma and A1367 might have moved during the age of the Universe ~1° on the plane of the sky as seen from Earth. Because Coma and A1367 are separated by more than 20°, the overall arrangement of the material in the supercluster cannot have been altered all that much over cosmic time. Of course, stars and galaxies form within the volume, but in a conventional sense the material we see in Coma has never mixed with the material we see in the cluster A1367. In other words, redshift surveys over wide angles on the sky generally reveal clues as to the primordial distribution of mass.

Gregory and I were successful at the 84-inch telescope in late April 1976, and after we finished our nighttime observing on the morning of May 1, Gregory took the spectra (a collection of small photographic plates) back to New York in order to measure the faint spectral features and to determine the galaxy redshifts. While I waited in Tucson for the results, I continued to work with the wide-field photographs we had taken with the 150-inch prime focus camera. My first job was to compile positions, galaxy morphological types, and shapes of cluster galaxies in Coma, A1367, and Hercules. Gregory came back to Tucson on July 14, 1976, ten days after the US Bicentennial celebration. We sat down in the library of the KPNO headquarters, and Gregory placed on the table in front of us several preliminary graphs including the 24° wide redshift survey map: redshift was on the vertical axis and right ascension (the east–west coordinate) was on the horizontal axis. In our plots, for convenience, the north–south coordinate is collapsed into the plane of the diagram. Although we could see large empty regions of space beyond the Local Supercluster, regions that contained no galaxies, I was not too happy that the plot was rectangular. To obtain reasonable proportions over the entire volume, it should have been a polar plot (with the Earth at the origin). I emphasized this, and even though Gregory grumbled and was impatient with the suggestion, I insisted that we change it. The galaxy coordinates and redshifts were on IBM punch cards, so I took the pack of cards to the central computer in the middle of the KPNO building, wrote a quick routine, and reduced the horizontal axis proportionately as the redshift approached zero. Instead of converting the rectangle into a cone diagram, we had a triangular one. It was with this triangular redshift map that we made our first attempt to understand the major structures that appear in the nearby Universe.

Redshift diagrams, even when plotted in a cone or triangle, do not show pure 3D structure. As mentioned above, galaxies travel through space, and they respond to the gravitational pull of their neighbors. For example, galaxies situated on the near side of a cluster (consider galaxies along the line of sight to the cluster center) will be accelerated away from us and will begin to fall rapidly toward their host cluster. This particular galaxy's apparent redshift, as encoded in its spectrum, will be higher than the cluster average. The converse is true for cluster members situated on the far side of the cluster center. The net result is a "Finger of God": a rich galaxy cluster in a redshift map assumes a cigar shape pointing toward our position at the lower apex of the redshift map. The center of the cigar sits at the appropriate 3D location of the cluster. The more massive the cluster, the more extended the "Finger of God." Examples of this effect can be seen clearly in Figure 5.4b, and in an attempt to eliminate this visual distortion, our first step was to circle the galaxy groups and galaxy clusters with ovals. We see in Figure 5.4a the raw data: each plotted point represents a single galaxy in our redshift survey. In the foreground (closest to the lower apex of the diagram) are numerous galaxies that belong to our Local Supercluster. Once we move beyond 60 million light-years (20 Mpc), we are looking into the deeper Universe that no one had explored before us in 3D. Both the Coma cluster and the A1367 cluster are circled and labeled in Figure 5.4b. Coma is at right ascension of 13h 00 m and A1367 at right ascension of 11h 45 m. Stretching horizontally between the two rich cluster cores, Gregory and I saw a bridge of galaxies connecting the two clusters. We anticipated this discovery in our observing proposal. Later studies would show that our bridge of galaxies is a small segment of the more extensive Great Wall that stretches far beyond our survey region, to both higher and lower right ascension (to the left- and right-hand edges of Figures 5.4a and 5.4b). We found that the density of galaxies in this supercluster bridge to be higher than the density of galaxies in and around the Milky Way.

In the volume between the Local Supercluster and the Coma/A1367 bridge, the galaxies are not uniformly distributed in space, and the most outstanding features in our redshift survey are two gigantic empty regions. These we immediately called "voids." The center of one void is located 150 million light-years (46 Mpc) from Earth, and the other is at a distance of 240 million light-years (75 Mpc). These enormous empty volumes were the first two cosmic voids ever identified. (One might argue that Herschel and Shapley saw the Local Void first,

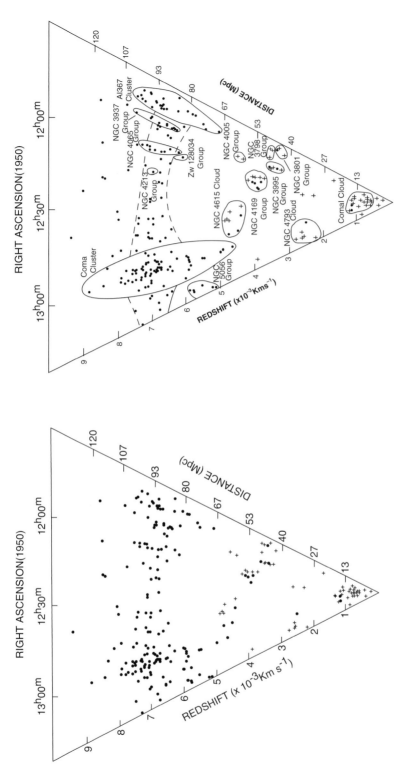

Figure 5.4a (left) 5.4b (right). Coma/A1367 redshift survey. This two-panel diagram is the main focus of the Gregory and Thompson 1978 cosmic voids discovery paper. The left panel by itself was already shown as Figure 2.2 where the conceptual beauty of redshift cone diagrams is described in the caption and the two cosmic voids are identified. The right-hand panel shows the lines we sketched in July 1976 when Gregory and I made our first attempt to make sense of the highly irregular galaxy distribution. We drew ovals around groups and clusters that displayed the "Finger of God" redshift-space distortion, and with dashed lines we designated the supercluster bridge connecting Coma to A1367. Our supercluster bridge is one segment of the larger "Great Wall" structure, later delineated and named by de Lapparent, Geller, and Huchra (1986). By permission of the A.A.S.: S. Gregory & L. Thompson (1978). *Astrophys. J.*, 222, 784–99.

but they did so on a 2D map of the sky.) These two voids turn out to be relatively normal in terms of the characteristic diameters of the multitude of voids known today. Even Herbert Rood would eventually give us credit for uncovering what he called a "hidden paradigm": the existence of large empty regions in the galaxy distribution that fill more than 60% of the volume of the Universe.

An important requirement of our observational program was to collect a complete sample of galaxy redshifts, so we selected a specific brightness limit and obtained spectra for all the brighter galaxies in our survey area. If the sample is incomplete, it might have been possible that we missed galaxies in a given direction and thereby obtained a misleading picture as to where the galaxies reside and where they do not. Actually, our sample is only "nearly" complete because there are a few low-surface-brightness galaxies (dwarf galaxies) whose redshifts were impossible for us to detect with the telescope on Kitt Peak. But it was safe for us to assume that these dwarfs are associated with galaxies we did detect.

We knew of only one other group who could have identified these huge empty regions and that was Chincarini and Rood. Of course, we had described to them, in a broad outline form, our wide-angle redshift survey plans in May of the previous year, and we essentially told them that our purpose was to map the 3D galaxy distribution. Because I had been busy working on our KPNO prime focus photographs of galaxy clusters, I had failed to look for their publications. I knew they planned to write two papers, one for rapid publication in the British journal *Nature* and the other for the *Astrophysical Journal*. I immediately got to work to find these publications in the KPNO library: Chincarini and Rood (1975, 1976). After finding each paper, I went through them carefully. At first glance, I was encouraged to see that neither showed a true 3D plot of the galaxy distribution.

From the start of their collaboration, Chincarini and Rood fell into the consistent habit of creating, for nearly every cluster paper they wrote, a rectangular redshift plot in which each galaxy is represented by a single point. The galaxy's radial distance on the plane of the sky – measured from the cluster center – was always on the horizontal axis, and the galaxy's redshift was on the vertical axis. This type of plot can mask 3D structure. For example, say there were two galaxies that happened to have the same redshift, but one galaxy was located east of the cluster center and the other west of the cluster center, both at the same radial distance from the cluster core. In the Chincarini and Rood plot, these two galaxies would sit next to each other and look like neighbors. With radial averaging relative to the cluster core, they simply could not see true 3D information. Furthermore, their plots were always rectangular: galaxies at the lower redshifts (in the near-field) are stretched across the bottom of the diagram like Greenland is stretched in some projected maps of the Earth. (This is the distortion I eliminated from the Gregory and Thompson redshift survey map in July 1976 when I changed our map from a rectangle to a triangle.) Chincarini and Rood stated many times that they admired and were emulating the historically significant work on the Coma cluster by Mayall (1960). However, by continuing to use the Mayall plot, they passed by the opportunity to visualize 3D structures. Anyone wanting to investigate 3D structures would have immediately abandoned this method.

The second encouraging point – from the Gregory and Thompson perspective – is that *nowhere* in their two papers (nor in any earlier paper) did they use any geometric term like "hole" or "void" or "empty volume" when referring to the galaxy distribution. In the second of their two papers, Chincarini and Rood used the rarely applied phrase "segregation in redshifts." This term can have several meanings. In the context of galaxy redshifts, the term "segregation in redshifts" was introduced in Chincarini and Martins (1975), a brief paper that is somewhat difficult to read. It provides redshifts for and describes objects in and around a peculiar group of galaxies called Seyfert's Sextet (Arp 1973). Seyfert's Sextet was being discussed in the 1970s by Halton Arp as a prime example of an object that displays non-cosmological redshifts. The issue of non-cosmological redshifts is clearly discussed by Chincarini and Martins (1975).

Chincarini confirmed in a discussion we had (via email) in late 2019 that he was, in every sense, discussing a non-cosmological interpretation of redshifts when he introduced the words "segregation in redshifts." Chincarini related to me that he would have made the link between his term "segregation in redshifts" and non-cosmological redshifts in his Chincarini and Martins 1975 paper more direct, had he not been under the influence of G. de Vaucouleurs at that time, a scientist who never strayed from conventional physics. No one else but Chincarini was using the term in astronomy. If galaxy redshifts had actually turned out to be non-cosmological – as Arp and Tifft kept insisting – Chincarini (working as an empiricist) could have claimed to have seen that effect first, too. Only after Gregory and I introduced the 3D geometric concept of "voids" into our Coma/A1367 supercluster analysis and plotted 3D redshift maps did Chincarini link his so-called segregation in redshifts with spatial holes. This late-time clarification was made in a controversial 1978 paper in the journal *Nature* that will be discussed later (Chincarini 1978). Once again, I emphasize the point that in not a single Chincarini and Rood publication that they submitted prior to the Gregory and Thompson Coma/A1367 Supercluster paper in September 1977 was the 3D concept of a "hole" or "void" ever introduced or discussed. Instead, Chincarini had non-cosmological redshift ideas in mind.

Chincarini and Rood had adopted a working hypothesis that was quite different from Gregory's and mine. Their aim in collecting galaxy redshifts at Kitt Peak in 1975 was to trace the galaxy density in the Coma cluster outskirts, to the point where the cluster halo fades into the background of so-called field galaxies. Chincarini's thinking was based on the model supported by Hubble and Zwicky where clusters are enhancements in the general galaxy distribution, that is, positive density enhancements that reside in an otherwise homogeneously distributed population of field galaxies. But we know today that the actual situation is more complex. The Coma cluster is embedded in the cosmic web. Galaxies located east and west of the Coma core are part of the supercluster bridge (the Great Wall). In other radial directions (i.e., toward adjacent voids), background galaxies are absent: the density of galaxies dips lower than the mean. Without realizing it, Chincarini and Rood found themselves in the awkward position of trying to determine how the halo of the Coma cluster fades into structurally complex surroundings. They simply did not understand this point: nowhere in their papers is this assumption even discussed.

Gregory and I took our time to write our Coma/A1367 Supercluster paper for the *Astrophysical Journal*. We had other distractions. I moved from KPNO to the University of Nebraska, Lincoln, and Gregory moved from a small state college in upstate New York to another in Bowling Green, Ohio. Starting at a new university is a job in itself. Immediately after our meeting in the KPNO library in July 1976, Gregory volunteered to write the first draft of our new paper, but when I saw his first draft near the end of the summer, it seemed to me that much of what he had written had an encyclopedic style. I placed parts of his first draft in an appendix to our new paper and wrote the second draft in such a way as to place our work in a broader cosmological context. I emphasized as best I could a 3D graphical overview of the general Coma/A1367 volume of space. Step-by-step, we iterated to a satisfactory result. We both returned to Tucson in the summer of 1977 where we sat together to make the final corrections to the manuscript. Some figures were still in rough form, but when I returned to Lincoln in late August 1977, I got these small jobs done. In the paper's abstract we wrote:

> In front of the Coma/A1367 supercluster, we find eight distinct groups or clouds but no evidence for a significant number of isolated "field" galaxies. In addition, there are large regions of space with radii $r > 20\ h^{-1}$ Mpc where there appear to be no galaxies whatever.

And then in the main body of the paper we wrote in greater length:

> There are large regions of space with radii $>20\ h^{-1}$ Mpc which contain no detectable galaxies, groups, or clusters, giving an upper limit to the

detected mass density in these regions of density $<4 \times 10^{-34}$ g cm^{-3}.
A redshift survey now being done by Gregory, Thompson, and Tifft
(1978) which examines the supercluster surrounding A426 (Perseus),
A347, and A262 shows that there exist even larger voids than any found
in the present study.

It is an important challenge for any cosmological model to explain
the origin of these vast apparently empty regions of space. There are
two possibilities: (1) the regions are truly empty, or (2) the mass in these
regions is in some form other than galaxies. In the first case, severe
constraints will be placed on theories of galaxy formation because it
requires a careful (and perhaps impossible) choice of both omega
(present mass density/closure density) and the spectrum of initial
irregularities in order to grow such large density irregularities. If
the second case is correct, then matter might be present in the form of
faint galaxies, and an explanation would have to be sought for the
peculiar nature of the luminosity function. Alternatively, the material
might still be in its primordial gaseous form (either hot or cold neutral
hydrogen), and the physical state of this matter may be similar to that
discussed in a number of speculative papers (see Rees and Ostriker
1977). A search for radio radiation should be made in the direction of
the voids.

These two paragraphs represent a pivotal change in perspective. We use the
word "voids" twice as a noun. We are clearly describing empty 3D volumes.
Until this point in time, nowhere in the astronomy literature had the term
"voids" been used (at least in the context of the large-scale structure in the
galaxy distribution). Next, we made a reasonable attempt to place the phenomena that we had discovered in a proper astrophysical context. This stands in
stark contrast to the Chincarini and Rood empirical statement that they had
observed "segregation in redshifts" with its ambiguous and potential non-
cosmological redshift implications. Finally, our visualization with the "triangular" cone diagram cinched the discovery.

When we wrote the words "severe constraints will be placed on theories of
galaxy formation because it requires a careful (and perhaps impossible)
choice . . .," we were making an indirect reference to the hierarchical theory
of galaxy formation. As explained in detail in Chapter 7, the traditional hierarchical model was being confronted in this era with the Zeldovich "pancaking
theory," and it seemed that the galaxy distribution Gregory and I had uncovered
might be more easily explained in terms of the "pancaking theory." Gregory and
I wrote another separate paper soon afterward aiming to see whether

observations of the Coma cluster itself conformed in any way to the Zeldovich picture (Thompson & Gregory 1978).

Gregory and I had agreed to a pledge of silence starting at the time we saw the first 3D graphs of the Coma/A1367 supercluster in the summer of 1976. We would tell no one about what we had discovered. We knew the potential implications of our work, and we realized that if we remained silent, we would have the luxury of time to write our paper carefully without significant worries about competitors. We thought that few, if any, other astronomers could match our data set: they would need too much telescope time.

The summer of 1977 was coming to an end, and soon both of us would have to return home separately to prepare for the fall semester. Gregory and I were on-track to submit our paper for publication in early September. (It arrived at the *Astrophysical Journal* on September 7, 1977.) With everything in place, we were finally free to discuss our results with others. Gregory and I agreed that the very first person we needed to talk to was William Tifft. On a mid-August afternoon, we walked from the KPNO headquarters building across North Cherry Avenue to his office located on the second floor of the old brick Steward Observatory office building. Tifft was scheduled to depart in several weeks for a meeting organized by the International Astronomical Union (IAU) entitled "The Large Scale Structure of the Universe" IAU Symposium No. 79. The meeting was being held in Tallinn, Estonia, USSR, September 12–16, 1977. Tifft and Gregory had submitted their applications to attend the meeting sometime earlier, but only Tifft had received an invitation to attend. We described to Tifft the concept of cosmic voids as distinct physical objects sitting within our survey volume, and he agreed to highlight our publication at the conference and to introduce and openly discuss the new Gregory and Thompson concept of "voids." The three of us also discussed new results that were coming in for the Perseus supercluster redshift survey that was a Gregory-Thompson-Tifft collaboration. The first of these redshifts had been collected with the Steward Observatory 90-inch telescope during previous fall observing seasons, and after Tifft looked at these new redshifts in a preliminary way with knowledge of our work on Coma/A1367, the Perseus data also showed evidence for another large void that we decided to cite in our Coma/A1367 paper.

Only Tifft and Gregory (1978) were listed as authors on the Tallinn conference paper despite the fact that it focused on the new key results from the Gregory and Thompson Coma/A1367 Supercluster paper. I do not recall whether I was asked to be a coauthor of the Tallinn conference paper or not, but even if I had been asked, I would have declined. I had no intention of being a coauthor with Tifft on a conference paper because of the risk that he would insist on discussing non-cosmological redshift concepts in a free-form way at the meeting, thereby linking my name to a concept that I did not accept. As stated before, Gregory had no such objection and in the end, Tifft mentioned nothing in the conference presentation on his non-cosmological beliefs. I decided to accept the inevitable fate of facing Tifft's non-cosmological redshift concepts (but in a more controllable format) in the upcoming Perseus super-cluster paper. That publication would be the only one where my name would be associated with any aspect of non-cosmological redshifts. When I presented preliminary results for the Perseus supercluster at the Austin, Texas, meeting of the American Astronomical Society in January 1978 (Gregory, Thompson, & Tifft 1978), there was nothing said about this topic.

5.5   Hercules Supercluster Results

Chapter 6 deals with the general significance of IAU Symposium No. 79 to the study of cosmic voids, but there is one supplementary topic from that meeting that belongs in this chapter on the early Arizona redshift survey work. This topic is the tale of the group effort to investigate the Hercules supercluster. Keep in mind that Gregory and I kept our mouths sealed so that Tarenghi, Chincarini, and Rood were unaware of the Coma/A1367 results until they learned about it in Tallinn. As mentioned earlier, Tarenghi was the Hercules project leader, and by the late summer of 1977, everyone but Tarenghi had finished their appointed jobs, and the Hercules data set was nearly complete. All five Hercules supercluster authors were asked and we all agreed to have a paper on Hercules read by Chincarini at the Tallinn IAU Symposium in September 1977. That put sufficient pressure on Tarenghi to complete the redshift data reduction. Sometime in the summer of 1977, in Europe, Tarenghi gave to Chincarini the complete (but unfortunately still preliminary) data set, including the list of galaxy redshifts in the Hercules supercluster. In that version of the redshift list, a number of Tifft and Tarenghi galaxy redshifts were just eyeball estimates, but it was good enough for a less-than-formal conference presentation. Here, I describe the series of events that happened next.

Because I did not attend the IAU Symposium, once the conference was finished, I was most interested to get a copy of the brief paper that Chincarini had prepared. The preliminary Hercules redshift data had not been circulated to the rest of the Hercules supercluster collaboration (in April 1978 we were scheduled to see the final version in Norman, Oklahoma). I simply wanted to know if there were other giant voids sitting in the foreground of the Hercules supercluster that might confirm the Coma/A1367 supercluster study. I cannot recall when I first saw our IAU conference paper on Hercules, but when I did,

I noticed that it contained no figure or graph displaying the Hercules galaxy redshifts. In fact, no trace of the actual data appears in the Tallinn conference proceedings. I saw this as somewhat vexing. After each paper was presented at the conference in Tallinn, a discussion was held and was entered into the conference record. Based on the transcript of this discussion, I could tell that Chincarini did display our redshift distribution during the meeting. But in the conference paper, the redshift data were nowhere to be found. When I asked directly about the data, Chincarini hesitated and then made (what appeared to me to be) an excuse by saying that his slide (showing the redshift plot) was of such poor quality that it could not be placed in the published conference paper. In the end, I had to wait to see confirmed evidence for voids until our Hercules supercluster group meeting in late April 1978 at the University of Oklahoma, at which time Tarenghi would provide to the group the final data set. Only then would I see that there was, indeed, a large cosmic void situated in the Hercules supercluster foreground.

It took many years before I figured out what Chincarini had done. This was confirmed in 2013, when Chincarini posted online his version of the presentation he had made the previous year at the 13th Marcel Grossmann Conference in Stockholm, Sweden (Chincarini 2013). In this presentation, Chincarini highlights a 1978 article he had published in the journal *Nature* entitled *Clumpy Structure of the Universe and General Field* (Chincarini 1978). Chincarini states in his Marcel Grossmann lecture that, because of this paper, he was the scientist who published "the first statistical evidence of the voids." He even talks in the Marcel Grossmann lecture about his supposed role in selecting the word "voids" over "holes" for the large empty regions. There is no basis for either claim. I explain why.

Chincarini's 1978 *Nature* paper contains the Hercules supercluster redshift data that "went missing" from the 1977 IAU Symposium presentation in Tallinn even though this redshift data set did not belong to him. It belonged to the five scientists in the Hercules collaboration: Tarenghi, Tifft, Chincarini, Rood, and Thompson (1979, 1980). What is potentially even more awkward for Chincarini is a possible link between Chincarini and the referee for the Gregory and Thompson Coma/A1367 Supercluster paper.[1] If so, in the period while Chincarini was writing his 1978 *Nature* paper immediately after the Tallinn meeting, he is likely to have had access to the pre-published Coma/A1367 manuscript that described Gregory's and my physical interpretation of cosmic voids. Chincarini submitted his paper to *Nature* with the Hercules redshift data in December 1977, three months after the IAU Symposium in Tallinn and three months after we submitted our Coma/A1367 manuscript to the *Astrophysical Journal*. Our Coma/A1367 manuscript was approved by the journal editor in

December 1977. In every sense, the basis for Chincarini's *Nature* paper is dubious: the data came from our five-scientist Hercules collaboration, and they were not his to publish. Before publishing his paper in *Nature*, Chincarini made no attempt to inform our group leader Tarenghi (nor me) as to what he was doing. His use of the word "voids" was adopted from the soon-to-be-published work of Gregory and Thompson. As for Chincarini's claim to have had a role in introducing the word "voids" into the astronomical literature, this is certainly incorrect.[2] Fortunately for me and Gregory, the date of a scientific discovery is tied to the date when a paper containing the discovery result is submitted to a journal and not when the paper appears in print or when it is discussed in a scientific meeting. An exception is required if the authors submit a paper and then significantly revise it during the review process (i.e., if they submit a revised version with altered content). In this case, it is the date of the new revision that is significant. Gregory and I submitted our Coma/A1367 Supercluster paper to the *Astrophysical Journal* in early September 1977, and we made only very minor revisions requested by the referee. Chincarini's dubious paper to *Nature* was submitted in December 1977 but it happened to appear in print before our Coma/A1367 Supercluster paper only because the printing and production time for *Nature* is especially short.

In the Marcel Grossmann lecture, Chincarini reveals that he invited Rood to be a coauthor on the 1978 *Nature* paper, but Rood declined (Chincarini 2013 p. 14 footnote "ee"). Rood's refusal is completely consistent with his life-long straight-arrow judgment. Who would publish a scientific result based on data they had taken from their own collaborators? This could have been handled properly in only two ways: our Hercules group redshift data and Chincarini's mathematical calculation based on that data could have been held for inclusion in the main Hercules supercluster paper by the collaboration, or alternatively, all authors names could have been added before Chincarini's paper was submitted to *Nature*. If all the authors in the collaboration had agreed, that would have been fair enough. Simply stated, Chincarini's result was not his to publish.

Some might wonder why all six Arizona collaborators – Chincarini, Gregory, Rood, Tarenghi, Thompson, and Tifft (listed here in alphabetical order) – were not part of a general discovery paper discussing cosmic voids for the first time. This is easy to answer. In July 1976, when Gregory and I first saw definitive evidence for cosmic voids in the Coma/A1367 redshift survey, we felt there was a risk to releasing our result, especially to Chincarini and Rood. They were in a position to move quickly to publish our concept of cosmic voids by reinterpreting their own 1976 data set. If they followed that path, we could have been excluded entirely. Our interactions with Tifft and Tarenghi were perfectly smooth. Tifft was a gentleman who never made a single inappropriate move, and Tarenghi was my good friend. However, our first meeting with Chincarini and Rood in the Kitt Peak cafeteria (14 months before we first saw the cosmic voids in our 3D maps) was somewhat tense and gave us a feeling of potential friction. Gregory had not been invited into the Hercules supercluster collaboration, so there was never any unity among all six scientists. I viewed the Hercules supercluster collaboration as an effort among equals led by Tarenghi, but Chincarini somehow got the idea that he was in charge. Gregory and I had already agreed in mid-1975 to release superb copies of our 4-meter telescope cluster imaging survey (compliments of reproductions made by the KPNO photo lab) to all of these scientists, and that seemed to me and Gregory to have been generous enough at that early stage. The Chincarini *Nature* incident described here confirms that Gregory and I made the correct decision to keep our discovery quiet until our Coma/A1367 supercluster manuscript was submitted for publication.

In late April 1978, we held a Hercules supercluster group meeting in Norman, Oklahoma. The complete Hercules data set was distributed to everyone in the group at that time. Tarenghi had completed the final redshift measurements (replacing the eyeball estimates) and we needed to get the paper written. Tarenghi came from Europe, Rood from Michigan State, and I drove down from Nebraska. Tifft was unable to attend. On Friday morning, April 28, I went to Chincarini's office in the Physics and Astronomy Department at the University of Oklahoma. By the time I arrived, Tarenghi and Rood were already working at a large table in Chincarini's office.

The Hercules data set was very respectable with nearly 200 galaxy redshifts. Half were from Tarenghi and Tifft, 25% were from Chincarini and Rood, and the remaining were measured at earlier times and were found in published catalogues. Before the April 28, 1978 meeting, no one had plotted the Hercules redshift data in a cone diagram: Chincarini presented the data at Tallinn (and in *Nature*) simply as a redshift histogram with the number of galaxies plotted in bins as a function of the observed redshift. I made a rough sketch of the cone diagram at the Oklahoma meeting and volunteered to have a draftsman make the final copy once I returned to Lincoln. Figure 5.5 shows this diagram. Sure enough, yet another enormous void sits in the supercluster foreground. The text of the paper itself became a "committee affair," in the sense that each one of us had separate aims. Rood wanted a virial analysis for each cluster in the sample (the measured positions and velocities of all member galaxies are summed in such a way to yield the total mass of the cluster). I wanted to see the cone diagram and to check for galaxy intrinsic alignment effects among the galaxies. Chincarini wanted to calculate the density limit for field galaxies. Tifft wanted to include a non-cosmological analysis of the galaxy redshifts. By the time we

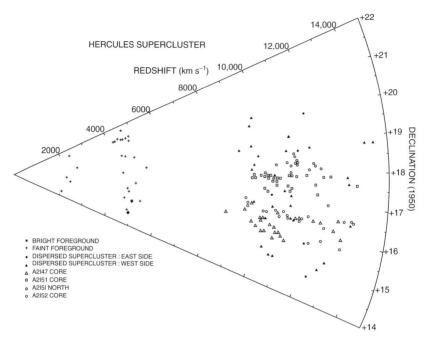

Figure 5.5 Hercules Supercluster redshift survey. This diagram was originally published in 1979 in the first of the two papers on the Hercules complex by Tarenghi et al. 1979. Each small symbol represents a single galaxy. Notice the dramatic 50 Mpc (160 million light-year) gap in the galaxy distribution in the immediate foreground of the Hercules supercluster (the supercluster is the collection of points between 9,000 km/s and 13,000 km/s). The survey volume is relatively narrow, spanning only 8° in the north-south direction, but this angle is exaggerated in the cone diagram by a factor of ~6 times. In this orientation with the apex of the cone on the left, the deeper Universe stretches off to the right. By permission of the A.A.S.: M. Tarenghi, W. Tifft, G. Chincarini, H. Rood & L. Thompson (1979). *Astrophys. J.*, 234, 793–801.

were finished with our three-day group meeting, the paper seemed to be coming together even though it was not elegantly written.

Our Hercules supercluster manuscript was submitted to the *Astrophysical Journal* in November 1978, after everyone had a chance to read and make corrections from the contributions of all coauthors. During the review process at the journal, the referee, among other smaller things, strongly objected to the section by Tifft on non-cosmological redshifts. Tifft reached an impasse with the referee, so the editor of the journal appointed George Abell to be the arbiter. After hearing all sides, Abell recommended that the paper be split into two parts so that the new observations could be published on their own with no further delay. The names of all five authors remained on the first paper. Abell then recommended that the data analysis be presented by itself and that the non-cosmological discussion be dropped. At that point, Tifft chose to remove his name from the second paper, and it was published with the remaining four authors: Tarenghi, Chincarini, Rood, and Thompson. Tarenghi remained the lead author on both papers and through the entire process maintained the lead role in the collaboration.

By the summer of 1979, the Hercules supercluster paper was complete, and its cone diagram confirmed the void and supercluster structure, first seen in the Coma/A1367 region (and in the early Perseus supercluster analysis). While Gregory and I applied for and received more telescope time at KPNO for a redshift survey of the double cluster A2179 and A2199, a likely candidate for yet another supercluster system, Chincarini, Rood, and I began to study the region between these two supercluster systems. By 1981, we published our analysis under the title *Supercluster Bridge between Groups of Galaxy Clusters* (Chincarini, Rood, & Thompson 1981), where we present evidence for an enhanced galaxy density across the 20° span that separates these systems. The discovery of cosmic voids was on a solid foundation, and we were moving beyond that accomplishment and into the investigation of broader and more extended structures.

In Tucson, on our way home after collecting spectra for A2197 and A2199, Gregory and I stopped to discuss our research with the retired Director of Steward Observatory, Professor Emeritus Bart Bok. The date was June 29, 1979. Bok had been given a luxurious office suite on the upper floors of the old 36-inch telescope dome adjacent to the Steward Observatory main building. Recall that he was Head of the Astronomy Department when Gregory and I started graduate school, so we confided in him and described our problem: our wide-angle redshift surveys were revealing supercluster connections between rich clusters punctuated by cosmic voids, and we suggested to him that our work was likely to be historically significant. We needed suggestions as to what we might do next to consolidate our position and to get word of our discovery to a broader audience. After a brief pause, Bart Bok said in his thick Dutch accent, "Well boys, I know exactly what to do. The editors of *Scientific American* are my good friends, and I will see to it that they publish a paper describing your work. Once something appears in *Scientific American*, the discovery priority is set."

We were pleased with Bok's suggestion and his support. We thanked him, and Gregory and I walked away satisfied that this matter was in good order. However, as time passed, it would become clear that Bok's suggestion fell far short of what was needed. We wrote our *Scientific American* article during the late summer and fall of 1979, as I was moving from Nebraska to Hawaii, and by the time of the 1979 Christmas holiday, Gregory and I were essentially finished with the manuscript. We sent it to *Scientific American* in early 1980, and then we began to wait for their reply. After several months passed, I telephoned the *Scientific American* editorial office. The editors apologized and said that they liked our contribution, but their backlog of articles was large. They would need to get back to us as soon as they could move forward to publish our article. This was the prelude to what Gregory and I would eventually see as a major setback and complication.

At a slow pace, Gregory, Tifft, and I were finishing the Perseus supercluster study as shown in Figure 5.6. At times, this same object had been called the Pisces-Perseus supercluster because it extends across both constellations (Gregory, Thompson, & Tifft 1981). It is one of the longest filamentary structures in the nearby Universe containing numerous galaxies and galaxy clusters. It extends more than 40° across the sky and sits at a distance of 220 million light-years (72 Mpc), making it somewhat closer than the Coma/A1367 supercluster. Early on, we had come far enough along in our Perseus study to have made reference to it in the original Coma/A1367 supercluster study. With preliminary data from 1977 we could tell that another large void sits in the foreground of this supercluster.

I presented a preliminary report on our Perseus supercluster study at the Austin, Texas, meeting of the American Astronomical Society in January 1978, a few months before the Hercules supercluster meeting convened in Norman, Oklahoma. The Perseus manuscript was refereed at the *Astrophysical Journal* through most of 1980, and it appeared in the journal in January 1981. Several years later, two widely cited papers, one by Zeldovich, Einasto, and Shandarin (1982) and the other by de Lapparent, Geller, and Huchra (1986), chose to reference only our Perseus supercluster analysis from 1981 and to ignore the 1978 Gregory & Thompson Coma/A1367 Supercluster cosmic void discovery paper. This is unfortunate because astronomers who read these two important papers (by Zeldovich et al. and de Lapparent et al.) were given the impression that the cosmic void discovery date is 1981 rather than 1978.

5.6   The Next Round: CfA1 Survey and the Boötes Void

A careful reader will have noticed that all spectra discussed up to this point in the book were recorded on photographic plates (after amplification by the image intensifier). Continuous advances in electronics meant that photographic plates would soon be replaced with electronic detectors. By the mid-

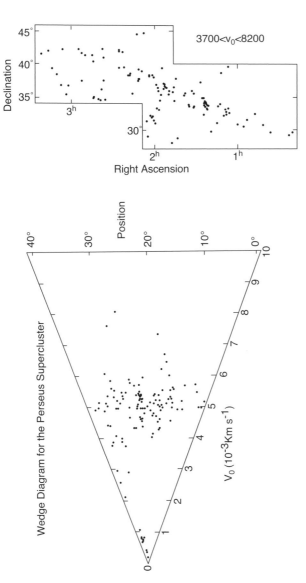

Figures 5.6a & b Perseus Supercluster redshift survey. This cone diagram from Gregory, Thompson, and Tifft (1981) is shown with its natural opening angle of 42° matching the angle on the sky. As originally stated in the Gregory and Thompson Coma/A1367 Supercluster paper, the cosmic void in the foreground of the Perseus Supercluster provided confirming evidence (from the southern galactic hemisphere) that cosmic voids are a common phenomenon in the nearby Universe. The top panel shows the spatial distribution (projected onto the sky) of the galaxies that are found in the supercluster redshift range, demonstrating the tight filamentary nature of this supercluster. Our early-galaxy selection (including many elliptical and S0's) shows an especially tight filamentary structure. By permission of the A.A.S.: S. Gregory, L. Thompson, & W. Tifft (1981). *Astrophys. J.*, 243, pp. 411–26.

1970s, the first experimental electronic readout devices were being tested at Steward Observatory and elsewhere. The most successful program in this regard was led by Stephen Shectman, then at the University of Michigan. Shectman, who later moved to Carnegie Observatories in Pasadena, California, was generous with his electronics design and willingly shared it with others. One group that acquired the Shectman camera design was led by Marc Davis, then at Harvard University. Davis aimed to start an extensive galaxy redshift survey with a dedicated 1.5-m (60-inch) diameter telescope located on Mt. Hopkins just outside of Tucson, Arizona. Davis assembled a research group at the Harvard-Smithsonian Center for Astrophysics (CfA) that included John Huchra, David Latham (b. 1940), and John Tonry (b. 1951). By late February 1978, the CfA system at Mt. Hopkins had started to collect image-intensified but electronically recorded galaxy spectra on a nightly basis. After several months of nighttime observing were completed, Davis was invited to give a scientific talk to describe his program as part of the joint KPNO-Steward Observatory weekly colloquium series.

Even though I was teaching at the University of Nebraska that fall semester, I happened to have been in Tucson using one of the KPNO telescopes when his talk was presented on October 26, 1978. At his talk, Davis only briefly described the Mt. Hopkins telescope and its spectrograph, and spent nearly all of his time explaining, in mathematical terms, how and why he considered it important to measure the mean mass density of the local Universe. The colloquium room was crowded for his talk, and there was a certain buzz in the air. Time was limited for questions after the talk was over, and I was unable to ask Davis publicly about what his redshift survey results revealed. I walked across North Cherry Avenue to my visiting astronomer office, and on my way out of the building that evening, I meet Marc Davis face-to-face walking down the hall.

The Coma/A1367 Supercluster paper was not widely available even by October 1978. It had appeared officially in the June 15 issue of the *Astrophysical Journal*, but in those days, the journals arrived at libraries from the printers up to six months late. Neither Gregory nor I had sent a preprint copy to anyone at Harvard. From our own perspective, the discovery of cosmic voids was on a strong foundation. We had confirming observations that showed large voids in the Hercules supercluster map and further confirmation from the soon-to-be published Perseus supercluster redshift map (Figures. 5.5 and 5.6). When I stopped Davis in the hall that evening, I was probing to see what he knew about the galaxy distribution. I said: "I had no chance to ask this question after your talk, but are you able to see any structure in the galaxy distribution in your redshift survey maps?" He replied in a somewhat dismissive tone with almost total emphasis on the first word: "WE are measuring the mean mass density of the Universe and not searching for structure in the galaxy distribution." He seemed impatient to move on, so I replied "Well, I enjoyed your talk," and I stepped aside so Davis could proceed on his way. In a matter of thirty seconds, our interaction was over, and I had my answer: Davis had no idea what opportunity he was missing: he was not even looking for structure that might be revealed by the galaxy distribution. As it turns out, the ongoing CfA1 redshift survey at Mt. Hopkins was "broad but shallow," while our supercluster redshift surveys were "narrow but deep." To see significant structure in the galaxy distribution as distinct features, a redshift survey had to include many of the fainter objects that Davis was not observing. What is now called the "First CfA Survey" (CfA1) collected 2,400 redshifts for only the brighter galaxies over a large area of the sky. In 1982, this group published their analysis of the large-scale distribution of galaxies, cited our "narrow but deep" surveys, and confirmed the existence of cosmic voids and a filamentary galaxy distribution (Davis, Huchra, Latham, & Tonry 1982). It was a necessity to cite our work because, on the basis of their "broad but shallow" survey, their results were not visually striking. The cosmic voids and supercluster interconnections were present but not in any way distinct in the CfA1 redshift survey. The last research papers from the CfA1 survey would be published by 1983. Both Davis (1988) and Huchra (2002) described details of their redshift survey when they sat for interviews for the American Institute of Physics.

In 1977, three other astronomers stepped forward to measure the mean mass density of the local Universe by doing their own modest redshift survey. Just like Davis, these three astronomers were Gregory's and my contemporaries. All three had held appointments at KPNO and had worked in the building on North Cherry Avenue at one time or another: Robert Kirshner (b. 1949), Augustus Oemler, Jr. (b. 1945), and Paul Shechter (b. 1948). I knew them well. Their observations of galaxies (including both redshifts and photometric brightness) were being collected in eight survey fields: four widely separated small areas in the North Polar Region of the Milky Way galaxy and four widely separated small areas in the South Polar Region. Kirshner, Oemler, and Schechter were familiar with our Coma/A1367 study. They acknowledged it and kept an eye out for the effects of the irregular redshift distribution of galaxies. They detected irregularities in the distribution of galaxy redshifts but left it as an unresolved issue at first (Kirshner, Oemler, & Schechter 1978, 1979).

After their first two papers were published, they began to extend their analysis to fainter galaxies and added to their collaboration the excellent spectroscopist and instrument builder Steve Shectman. At this point, the group of four (see Figure. 5.7) noticed a significant gap in the galaxy redshift distribution in three of their four northern fields. The gap stretched in depth over a distance range of 300 million light-years and the three fields were each separated on the sky by 35° (about 370 million light-years at that distance). They reasoned that if the entire region between their three sample fields is empty, they had evidence for the discovery of an unprecedentedly large void. Their paper describing these results became known as the *Million Cubic Megaparsec Boötes Void* (Kirshner, Oemler, Shechter, & Shectman 1981).

The Boötes void was a significant discovery because it represents an extreme example of the cosmic void phenomenon in our vicinity of the Universe. The huge size of this void was used repeatedly to test which cosmological model provides the best fit to the observations. However, their published paper was only borderline-fair. It begins with broad sweeping cosmological statements and no reference to earlier work. The paper eventually introduces our published cosmic voids – Coma/A1367, Hercules, and Perseus – and analyzes our results side-by-side with their new Boötes void, but the comparison is done deep within their paper. Even a careful reader has to look hard to find references to the original cosmic void discovery work. Then, rather than waiting to gain community acceptance of their work through scientific channels in the way Gregory and I had done, this group hired a public relations firm out of Yale University to disseminate information to news outlets around the world.[3] In this regard, the members of this group were also being pioneers. For example, when we asked Bart Bok how we might gain wider acceptance for our discovery of cosmic voids and the "bridges" we had detected between major galaxy clusters, hiring a public relations (PR) firm was not on his list of suggestions. This was not a normal path for astronomers in the mid-1970s. In fact, we had been instructed very clearly by our Arizona professors during graduate school that scientists who publicly promote their own work were stepping out of bounds.

But for the Boötes void group, the PR worked unbelievably well. Their original paper was cleverly written and their PR campaign so successful that many astronomers immediately gave exclusive credit to this group for the discovery of cosmic voids. I personally watched the Walter Cronkite Evening News in early October 1981 from my apartment in Honolulu to see how this "giant hole in the Universe" was described. Was it too much to have expected the simple phrase: "Although other holes have been found, this is by far the largest ... "? Of course, nothing of the kind was mentioned in the broadcast report. Newspapers across the globe reported the story in the same manner: here is a unique one-of-a-kind hole in the Universe, implying (but not stating) that Kirshner and his collaborators had truly discovered the cosmic void phenomenon. In Chapter 8, I provide a timeline to compare the discovery of cosmic voids with the discovery of the CMB, and I review how the CMB became known to the general public. As a preview to Chapter 8, here I pose the following rhetorical question. What if Penzias and Wilson had not made public news for their discovery of the CMB in 1965, but later Ned Conklin, who discovered the motion of the Milky Way relative to the rest frame of the CMB radiation, had announced his work through a hired public relations firm, never stating but simply implying that he had discovered the CMB itself?

Equally significant to Gregory and to me, *Scientific American*, after making us wait in the wings for 18 months, decided on the day after the CBS news broadcast to contact us. They wanted a completed and updated article as soon as possible. But at the same time, *Scientific American* gave precedence to the Boötes void group. An article in *Scientific American* describing the Boötes void appeared in the February 1982 issue, and the Gregory and Thompson article in the March 1982 issue. So much for Bart Bok's idea as to how we could best establish our priority in the broader astronomy community for our discoveries! Bok kept us anchored in the conservative style of the early twentieth century, while those around us were moving forward rapidly.

Figure 5.7 Boötes void consortium. The four astronomers shown here are the coauthors of the paper entitled *A Million Cubic Megaparsec Void in Boötes* published in the *Astrophysical Journal Letters* in September 1981. The small inset image shows the group in 1979 at the time the Boötes void work was underway. The larger image shows the same scientists 34 years later in 2013. In the small inset image (left to right): A. Oemler, Jr., R. Kirshner (in front), S. Shectman, and P. Schechter. In the larger image (left to right): R. Kirshner, A. Oemler, Jr., P. Schechter, and S. Shectman. Reproduced with permission: copyright for the larger image by Stephen Shectman and for the smaller inset image by Robert P. Kirshner.

5.7   1982 IAU Meetings and Zeldovich's Neutrino Dark Matter

The International Astronomical Union (IAU) is the one organization astronomers have to coordinate activities around the globe. The IAU General Assembly meets every three years. During these meetings, astronomers participate in Joint Discussions on many topics, ranging from planets and stars to galaxies and cosmology. The IAU also sponsors smaller specialized meetings called symposia. These are often clustered around the General Assembly. In 1982, the IAU General Assembly met in Patras, Greece, and IAU Symposium No. 104 was organized on the Island of Crete. The smaller meeting was entitled "Early Evolution of the Universe and Its Present Structure." As a member of the IAU, I attended both, and I arrived in Patras on August 17 and in Kolymbari, Crete, on August 29, 1982.

The Athens International Airport in late August was insufferably hot, and the bus arranged to carry IAU participants to Patras was not air conditioned. Neither were the dorm rooms at the University of Patras where the conference was held. After about 2 a.m. the heat subsided, and I managed to get some sleep in my assigned university dorm room. At morning breakfast in the university cafeteria, I sat down at a table of six or eight astronomers, one of whom was Robert Kirshner, the first author of the Boötes void paper. Soon University of Cambridge astronomer Sverre Aarseth (b. 1934) joined the group. Aarseth's contributions are described in Chapter 7. I knew I was in good company.

In 1982, the Cold War was in full swing, and IAU members learned at the beginning of the General Assembly that a dissident astronomer, who worked (among other things) on models of galaxy formation, had been blocked by the Soviet government from attending the General Assembly. Leonid Ozernoi (1939–2002), a political supporter of Andrei Sakharov (1921–1989), was reported to be on a hunger strike in Russia after being denied the right to travel to Patras. In this atmosphere, much to everyone's surprise, the widely acclaimed theoretical physicist Yakov Zeldovich appeared at the IAU General Assembly sometime on Monday morning of the second week. He had been invited to deliver an Invited Discourse. Zeldovich appeared to be unavailable much of the day on Monday. But that evening, August 23, 1982, he gave his lengthy invited talk entitled "Modern Cosmology" (Zeldovich 1982) in the ancient open-air Roman Odeon theater, a short bus ride from the University of Patras. It was my good fortune to have met him at that time – with a brief introduction and a handshake – and to have heard him speak. Most significant to me was how freely he used the word "voids" as well as the general flow of his ideas that integrated the concept of cosmic voids into an explanation of galaxy formation. When I reread his presentation some 35 years later, I can see how, in his quick and deep mind, he was able to synthesize a pioneering and broad picture of modern cosmology. Some of the ideas he discussed turned out to be incorrect. For example, in this 1982 address he speculated that the dark matter consists entirely of massive neutrinos, and that the large-scale structure forms first (he called them pancakes) and galaxies fragment from them, that is, top-down galaxy formation, as I describe in Chapters 2 and 7.

My trip to Greece was most memorable. There were bus tours organized by the IAU to historic sites like Delphi, Olympus, and the Tomb of Agamemnon. During these tours, I happened to fall into the company of several astronomers who would later contribute to studies of cosmic voids and supercluster structure, including Richard Kron (b. 1951) from the University of Chicago and J. Richard Gott, III (b. 1947) from Princeton University. In the 1990s, Kron would be one of the key forces behind the next-generation redshift survey funded by the Sloan Foundation. After the discovery of cosmic voids was firmly secured, Gott would lead an effort to understand the topology of the large-scale galaxy distribution. At one of the bus tour's refreshment stops on a blazingly hot late morning, we sat at a roadside cafe on the road to Delpi, and Gott acted out the general feelings of the entire group by personally ordering and drinking in quick succession six bottles of ice-cold Coca-Cola.

By August 29, 1982, I had arrived in Kolymbari on the Island of Crete with nearly 200 other astronomers and cosmologists to participate in IAU Symposium No. 104, "Early Evolution of the Universe and Its Present Structure." The four-day meeting started on August 30. My presentation was on the redshift survey Gregory and I had completed on the volume of space surrounding the cluster pair A2197 and A2199, where we found these two Abell clusters, again embedded in a common supercluster with cosmic voids in the foreground. The meeting was organized by Abell and Chincarini (1983), and it was held on the grounds of a monastery located on the north coast of Crete. In the time that had elapsed since the IAU Symposium held in Estonia (discussed in detail in the following chapter), the significance of redshift surveys, cosmic voids, and superclusters had become obvious to all but the most die-hard conservative cosmologists. All key redshift survey groups attended: I came to represent myself and Gregory, Huchra and Latham came to represent the Center for Astrophysics group at Harvard, Shectman came for the Boötes void group, Chincarini had teamed with Tarenghi to present a paper (by then Rood had withdrawn from his collaboration with Chincarini), and Ricardo Giovanelli was there to represent the 21-cm wavelength surveys from Arecibo Observatory in Puerto Rico. The explosive growth in redshift surveys was in direct proportion to their future significance to astrophysics and cosmology.

The opening keynote address at the IAU Symposium meeting was given by the venerable Jan Oort, who was by then in his early 80s. Most refreshing to me at the conference were four to five papers written by theorists who discussed cosmic voids. Ironically, and perfectly fitting to me, was Virginia Trimble's conference paper on the importance of supernova searches to cosmology. She was clear to say that she simply intended to spur others on and to help them acquire funds for this important endeavor. My NSF application for funds to support a cosmological supernova search had been turned down less than two years earlier. I was also delighted to meet Professor Elizabeth Scott from UC Berkeley who had worked on the imaginative analysis of the Shane and Wirtanen galaxy data with Jerzy Neyman in the 1950s. While no one officially summarized the conference, I recall the informal summary at the end of the meeting by Abell. He described the large-scale distribution of galaxies as being somewhat similar to a view of the Los Angeles city lights at night from his home on Mulholland Drive. The grid of street lights resembled the ridges of galaxies in filamentary superclusters, and cosmic voids were the dark areas in between. It was a fitting comparison.

Jan Oort's keynote address at the IAU symposium was a sparse skeleton when compared to his full review entitled "Superclusters" in the *Annual Review of Astronomy and Astrophysics* (Oort 1984). His full article is a fair and balanced scientific review of the early work. It includes a discussion of both observations as well as theoretical models. He highlighted the Arizona redshift survey work including the 3D cone diagrams of the Coma/A1367, Hercules, and Perseus superclusters. He presented evidence for supercluster structure from all other available sources too, including the Local Supercluster and the shallow but broad CfA1 Redshift Survey. When discussing the Perseus supercluster, he presented first the (rather weak) study from Tartu Observatory (Einasto, Joeveer, & Saar 1980) and then placed these early results side-by-side with the Gregory, Thompson, and Tifft (1981) Perseus redshift survey maps as well as the 21-cm radio wavelength Arecibo Observatory redshift survey maps (Giovanelli, Haynes, & Chincarini 1986).[4] Oort discussed voids, and even touched on the critical problem as to how such significant structures can grow in the galaxy distribution even though the CMB radiation is so smooth. Gregory and I first raised this issue in our Coma/A1367 Supercluster paper by saying that "it requires a careful (and perhaps impossible) choice of both omega (present mass density/closure density) and the spectrum of initial irregularities in order to grow such large density irregularities." Oort gave his own thoughts about this problem at the end of Section 10.1 in his review article by saying that "these facts argue in favor of the scenario in which superclusters formed first, and galaxies afterwards." This is the "top-down" model associated with work done by the Zeldovich group. The Zeldovich concept was competition for Peebles, who consistently stuck with the "bottom-up" or hierarchical model. This was just about the time that cold dark matter was beginning to be considered seriously as a component in the formation of both the supercluster structure and individual galaxies, something that is now a key feature of the LCDM model of the Universe.

As mentioned above, the First CfA Redshift Survey came to a close in 1983 as Davis, the group leader, left Harvard University to join the faculty at UC Berkeley. The spectrograph system that he and his group put into operation fell into the hands of the astronomers who were left behind at the Harvard-Smithsonian Center for Astrophysics. Under the guidance of John Huchra, a member of the First CfA Redshift Survey team, observations with the telescope and spectrograph system continued. Huchra described his own role in the early 1980s to be somewhat of a "service observer," where CfA scientists would make requests for observations of specific astronomical objects, and he would complete them. One of the beneficiaries was CfA astronomer Margaret Geller (b. 1947), who advised several successful PhD thesis students and relied on observations from the Mt. Hopkins redshift system to study clusters of galaxies, but the redshift survey style of observing temporarily came to a halt in the period 1983 to 1985.

5.8   CfA2 Survey and the "Great Wall" of Galaxies

The previous mode of observing was revived in about 1985 when the Second CfA Redshift Survey began (and continued thereafter an additional ~10 years). It was only when the CfA2 survey began that Geller became directly involved apart from having received service observations from Huchra (as described by Huchra 2002). The new CfA2 survey systematically collected redshifts for a sample of galaxies significantly fainter than those observed in the CfA1 survey. According to Huchra (2002), a discussion ensued as to the best survey strategy, and the group reviewed different options. The original Davis strategy was to cover the full sky (visible from Mt. Hopkins) but to observe only the brighter galaxies. Other choices were to conduct deep and complete sampling over smaller areas or, finally, to sample in a more complete fashion along strips across the sky. Huchra suggests that the practical realities of observing efficiency forced the decision to favor strips. It was his view that the decision was a practical one, not driven by some deeper scientific purpose. The first strip to be selected was 6° wide, approximately 117° long, and centered at about +32° north declination. This area happened to cut directly through a 20° long portion of the Gregory and Thompson Coma/A1367 supercluster survey that itself spanned 24° by 15°. Some have erroneously called our redshift survey work "pencil beam surveys," but our Coma/A1367 investigation covered ~360 square degrees, and our Perseus supercluster investigation covered 482 square degrees, whereas the first strip of the CfA2 survey (de Lapparent et al. 1986) spanned 702 square degrees. As discussed in Chapter 8 and summarized in Table 8.1, by 1984, our combined survey area was over 2,000 square degrees and exceeded that of the first CfA2 slice by nearly three times.

There was one interesting difference between the CfA1 and the CfA2 surveys. Davis began his work before we announced the discovery of cosmic voids in 1978, but by the mid-1980s, our work had become well known. Ironically, in the meantime, this group of astronomers at the CfA seem to have slipped under the protective wing of east coast theoretical cosmologists, including those who suggested that cosmic voids and the large-scale structure did not exist. Both Davis and Geller were former PhD students at Princeton University, the home institution of John Bahcall and P. J. E. Peebles, who were holdouts for a homogeneous galaxy distribution as described at the end of Chapter 4.

Gregory Bothun, now at the University of Oregon, was a Harvard-Smithsonian Research Fellow in the period 1981–1983 who worked side by side with Huchra and Geller, and he shared with me the reactions of CfA astronomers to the Arizona redshift surveys when our work appeared in the astronomy journals. He reported that Huchra and Geller read and discussed our redshift survey papers, but neither of them accepted nor believed what our data showed: a highly inhomogeneous distribution of galaxies containing distinct cosmic voids and bridges of galaxies between galaxy clusters. In an American Institute of Physics Oral History interview, Huchra states that he ascribed to statistical fluctuations, the observed irregularities that had been seen in the galaxy distribution prior to the mid-1980s. He thought they were statistical flukes. This is surprising, given that no more than five years earlier he published a paper showing that among relatively nearby galaxies, 99% were members of multiple systems or groups with only 1% of the galaxies being isolated (Huchra & Thuan 1977). If essentially all galaxies are in groups, how hard is it to recognize and accept cosmic voids? His views may have been influenced in the early 1980s by Davis and Geller and perhaps indirectly by Peebles.

Princeton Physics Professor P. J. E. Peebles plays a significant role in this story. Geller (1974) wrote her PhD thesis under Peebles' direction, and in her early graduate school years, Geller and Peebles spent extended periods talking to each other. Although Peebles was not Davis' thesis advisor (that was Professor David Wilkerson), Davis and Peebles had an early and rich collaboration. Peebles doubted the significance of any structure in the galaxy distribution – apart from rich clusters – and posed as an alternate hypothesis that the

filaments of galaxies reported in the large-scale structure are likely to be figments of the imagination "because the eye is so adept at finding patterns even in
noise." I happened to be at a scientific meeting in Trieste, Italy, in
September 1983, when the statement I quote above was made in
a presentation by Peebles (1983). At the time of his talk I was highly irritated
by these suggestions. As if the Boötes void publicity was not enough for Gregory
and me to deal with, now we had to deal with a world-renowned cosmologist
who doubted the reality of structure in 3D redshift maps that we took as
obvious.

It was fortunate that Jaan Oort also attended the meeting in Trieste; he was in
top form, even at age 83, having just written his excellent review article on
superclusters. During the question and answer period, Oort asked Peebles about
the very point that bothered me the most. Oort made the following clever
statement as reported at the end of Peebles' conference publication: "I disagree
with your implied conclusion that the supercluster features that have been
discussed may be just only chance configurations. As examples to the contrary,
I point to the Local Supercluster which can hardly be considered something that
has accidentally struck one's eye without having a physical reality, and to the
Coma supercluster which, in my opinion, is well isolated in the CfA position/
velocity plots." At that point, Peebles began to backtrack somewhat. This inter-
change settled my mind enough that I was able to sit through the remaining
question and answer period without making any comment of my own. These
examples show that, even five years after making a remarkable discovery,
Gregory and I still had work to do. Oort was convinced, but Peebles remained
a holdout.

Doubts about the reality of cosmic voids, prevalent primarily on the eastern seaboard of North America, disappeared quickly with the publication of
the first strip of the CfA2 survey. By the summer of 1985, the Geller and
Huchra group with the active participation of graduate student Valerie de
Lapparent had completed the data set for the first strip, and their paper was
submitted to the *Astrophysical Journal Letters* on November 12, 1985 (de
Lapparent, Geller, & Huchra 1986). The 3D redshift survey map from this
publication was already briefly described in Chapter 2. They called it "The
Slice of the Universe."

It was absolutely no news to me and Gregory that the local Universe is filled
with the structural features visible in The Slice of the Universe. We had seen
similar structure (and in some cases, the identical structure) in our own redshift
survey maps because in their central region the two maps overlapped as shown
in Figure 5.8. We described the extent of our all-sky coverage in *Scientific
American* in 1982, where we showed a composite 4π all-sky map displaying the

various regions across the sky that we had surveyed up to that point in time. Even so, the de Lapparent et al. result is important for two reasons. First, their map shows an east–west sweep that spans a contiguous 117°, whereas our longest (the Perseus supercluster) spans one-third of that angle. More significantly, the CfA2 study included fainter galaxies than Gregory and I observed. Even though the CfA2 data improved the view, the scientific results confirm our original discovery of cosmic voids, but they did add one significant new result. As mentioned in Chapter 2, when Geller and Huchra published the complete results from the CfA2 redshift survey (Geller & Huchra 1989), they made the point that the extent of the largest supercluster features in maps of the galaxy distribution appear to be limited only by the size of the survey. In the Chapter 8 timeline, the CfA2 de Lapparent et al. result is placed in its proper historical context as key evidence in a continuous series of structural discoveries that started with the original Gregory and Thompson 1978 discovery of cosmic voids and the bridge of galaxies that connects the rich cluster cores of Coma and A1367.

While Gregory and I openly acknowledge these achievements of the CfA2 redshift survey, the CfA2 group, including both Margaret Geller and John Huchra, generally did not reciprocate. Instead, they often described both in public statements and in scientific talks a story in which they were the discoverers of all aspects of the large-scale structure. Whenever they could manage to do so, Geller and Huchra cited Kirshner, Oemler, Schechter, and Shectman (1981) for the discovery of cosmic voids (i.e., the Boötes void). When asked once about the work by Gregory and Thompson, Geller put a puzzled look on her face and replied: "I thought I referenced that." The fact is, the de Lapparent, Geller, and Huchra (1986) paper referenced only the Gregory, Thompson, and Tifft (1981) Perseus supercluster study but made no effort to cite the cosmic voids discovery paper from 1978 by Gregory and Thompson.

When individual researchers and research groups exaggerate their own credit and diminish that of their competitors, it is the option of senior scientists who present keynote addresses at conferences, write review articles, and write monographs to tell a more balanced account. One respected scientist who accepted this responsibility is the late Allan Sandage (1987), who was invited to give the keynote address to open IAU Symposium No. 124 held in Beijing, China, more than a year after the CfA2 "Slice of the Universe" redshift map was published by de Lapparent, Geller, and Huchra. In his address, Sandage said the following about the discovery of the large-scale structure in the galaxy distribution:

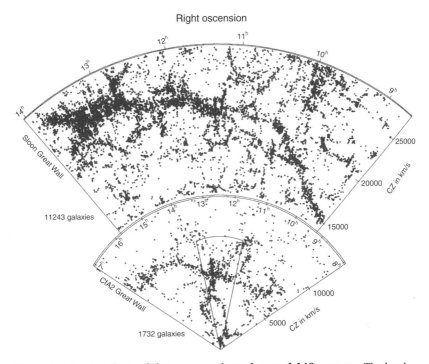

Figure 5.8 Comparison of three successive galaxy redshift surveys. The level of complexity revealed by galaxy redshift surveys can be traced in three major steps in this diagram that includes research spanning 25 years. The smallest fine-lined pie-shaped wedge in the center of the lower diagram represents the original redshift survey area of Gregory and Thompson from 1978. The entire lower 117° wide wedge represents the survey area of the de Lapparent et al. (1986) galaxy redshift survey as interpreted by Gott et al. (2005). Obviously, the two earliest surveys overlap in their central areas, although only the 1986 study is broad enough to capture the full beauty of the prominent structural feature called the Great Wall of galaxies. Even more striking is the Sloan Great Wall of galaxies that stands out with high contrast in the even deeper Sloan Digital Sky Survey area in the upper section of the diagram. The results from this series of three surveys clearly demonstrate the progression of observations that revealed the large-scale structure in the galaxy distribution. As Stephen Dick has suggested, the discovery was a process and not an event. By permission of the A.A.S.: J.R. Gott, et al. (2005). *Astrophys. J.*, 624, pp. 463–484.

[C]onfirmation from a different direction soon came using the addition of a third dimension redshift space. The convincing data and power visualization was by Gregory and Thompson (1978, their Fig. 2). This paper marks the discovery of voids, which have become central to the subject. Prior work by Einasto et al. (1980 with earlier references), Tifft and Gregory (1976), and Chincarini and Rood (1976), foreshadowed the development, but the Gregory and Thompson discovery is generally recognized as the most convincing early demonstration.

Geller also attended IAU Symposium No. 124 and discussed in her presentation the CfA2 "Slice of the Universe." She included in her published conference paper not a single reference to any of the early work of Joveer, Einasto, Chincarini, Rood, Gregory, Thompson, or Tifft.

6

Galaxy Mapping Attempt at Tartu Observatory

As Jaan Einasto describes in his extensive monograph (Einasto 2014) and in a shorter but more recent account (Einasto 2018), over a number of years in the 1960s and 1970s, he and his colleagues at Tartu Observatory met with and, at times, coordinated some portion of their research work with the theoretical astrophysics community in Moscow, a community that was home to well-known Soviet scientists including I. Shklovsky (1916–1985), Ya. Zeldovich, and Zeldovich's students A. Doroshkevich (b. 1937), I. Novikov (b. 1935), and R. Sunyaev. These scientists exchanged ideas at meetings within the USSR. Only occasionally could they afford or were they permitted by Soviet authorities to attend Western scientific meetings. Given the circumstances Einasto endured, his success story is extensive, remarkable, and admirable.

Prior to 1975, Einasto worked primarily on models that describe the structure of external galaxies like our neighbor, the Andromeda galaxy (M31), and the giant elliptical galaxy M87, which sits near the center of the Virgo cluster of galaxies. His galaxy models aimed to simultaneously fit the brightness and the velocity distributions of stars in each of these galaxies. By applying mathematical models to observations that had already been published in astronomy journals in the West, Einasto demonstrated the need for massive halos in the outer parts of galaxies: matter that was distributed in what he called, at that time, a corona. First, he suggested that galaxy halos might be composed of faint stars, but during discussions with others in the Soviet research community, they collectively concluded that galaxy halos are composed of dark matter. Because astronomical information was exchanged – with some difficulty – with scientists in the West, the work of Einasto and his Tartu Observatory colleagues was never too far out of sync with their Western counterparts, who also were becoming advocates of dark matter. But one aspect of Einasto's research was clear: he did not generate new observations. He and his Tartu Observatory colleagues seemed to have no access to modern observing facilities useful for studying galaxies and the Universe at large.

The conditions in the Soviet Union stood in stark contrast to the situation Gregory and I experienced in the early 1970s. During our graduate training, we were educated and intensively examined – to qualify for our PhD degrees – in many areas of physics, theoretical astrophysics, optics, and observational astronomy, but we did not live in the midst of nor exchange ideas directly with individuals like Zeldovich, Shklovsky, or Doroshkevich. Among the small number of astronomers making redshift observations from Arizona, few were well versed or well read on topics related to theories of galaxy formation. Because of the specialty I chose for my PhD thesis – the origin and nature of galaxy angular momentum – I was the best informed of the extended group of Arizona redshift survey astronomers in galaxy formation theories. Tifft was challenging the foundations of physics and took pride in working on an empirical basis. Gregory was not as radical but tended to be supportive of Tifft. Our redshift survey work and our study of general galaxy properties (galaxy morphology, ellipticity, and orientation) were open-minded, empirical, and exploratory. Most importantly, we had access to the best and newest technology on some of the larger telescopes in the world. Determining the nature of the large-scale 3D galaxy distribution required complete galaxy redshift survey samples, and we were in a position to generate these new data sets.

6.1   Tartu Observatory 1.5-m Telescope

Einasto describes in both of his personal accounts that Tartu Observatory, under his direction, built a new 1.5-m telescope for their observatory (Einasto 2014, pp. 71–4; Einasto 2018, pp. 32–3). The telescope was completed in 1975, at the height of the earliest searches for supercluster structure. The telescope was equipped with a spectrograph, and although Einasto does not discuss the detailed spectrograph design, he states that it was built to obtain spectra of stars. Then, he says that in order to obtain spectra of galaxies they ordered a nitrogen-cooled optical multichannel analyzer as a detector.[1] This was a major expense, especially in the 1970s, for an organization located within the USSR. Even so, not a single spectrum of a galaxy was published from this system in exactly the time frame when galaxy spectra provided the crucial path to discovering cosmic voids and large-scale structure in the galaxy distribution. Instead, Einasto and his colleagues ended up scouring all available scientific papers and catalogues, trying to find galaxy redshifts published by other astronomers. Regarding his own observatory, Einasto states that "the rate of spectra collected was low due to the small number of clear nights in our climate."

Jaan Einasto (b. 1929)

Jaan Einasto is an astronomer and Estonian patriot who experienced and endured the transformative changes of the political landscape in the Baltic States, associated with World War II and the Cold War. As a member of Tartu Observatory and the Estonian Academy of Sciences, he was a driving force in support of astronomy research in his home country. His father changed the family name to Einasto, a rearrangement of the letters in the country's name Estonia. Jaan Einasto's entire life has been associated with the Tartu region: he was born, educated, and became a researcher and a professor in the Tartu region of Estonia. Einasto studied physics and mathematics at the University of Tartu and earned his PhD in 1955, and his senior research doctorate in 1972. As described in this chapter, he studied the structure of individual galaxies and did pioneering work in both dark matter and the large-scale structure in the galaxy distribution, and has published over 190 scientific papers. He continued, over many years, to study the fundamental properties of the large-scale galaxy distribution. When Zeldovich presented the Tartu astronomers with a golden opportunity to participate in 1974 in answering fundamental questions about galaxy formation, Einasto and his colleagues were well aware of Zeldovich's scientific reputation as well as his stature in the authoritarian Soviet system. In 2014, Einasto published the monograph entitled "The Dark Matter and Cosmic Web Story." Biographical information is from Einasto (2014, 2018). Photo reproduced with permission: copyright Jaan Einasto.

One of my own areas of expertise is optical design, and in 2012, I analyzed the technical properties of the spectrographs used by Slipher and Humason to obtain the best possible galaxy spectra in the early 20th century (Thompson 2013). I wanted to pinpoint the reasons for the early success in measuring galaxy redshifts at Lowell and at Mt. Wilson Observatories. The key technical issue, as deduced for the first time by Slipher himself, is that spectrographs designed for stars would not work for galaxies. Slipher came to this conclusion based on experience at the telescope. He was put in charge of a spectrograph designed to study stars, and he redesigned it for his work on galaxies. The entrance slits on the highly successful nebular spectrographs used by Slipher and Humason were very wide (6–8 arcseconds), much larger than any entrance slit used for the spectra of stars. Also of key importance is the focal ratio (f/ratio) of the final camera lens in the spectrograph. To make a reliable redshift measurement, the camera lens needs to reduce the size of the image of the galaxy's spectrum (as it comes through the very wide spectrograph slit) to a tiny patch of light to match the spatial resolution of the detector. The telescope diameter is not a significant factor because galaxies are extended objects. Slipher used a 24-inch diameter telescope to measure his pioneering galaxy spectra. Humason used the famous 100-inch diameter telescope at Mt. Wilson Observatory for his work, but success came for Humason and Hubble, not because of the large telescope aperture, but from the excellent state-of-the-art spectrograph optics provided to them by the technical staff at Mt. Wilson Observatory. Einasto's Tartu Observatory 60-inch (1.5-m) telescope should have worked quite well for galaxy spectroscopy if it had a well-designed nebular spectrograph – not one designed to detect stars – and the right detector. The early redshift surveys at the Center for Astrophysics (CfA1 and CfA2) were completed on a telescope the same size as the telescope Einasto built in the 1970s.

Climatic conditions play a role, but other scientists have made bold advances in cosmology from inferior locations. Consider the historically significant work of William and Caroline Herschel described in Chapter 3. Herschel observed the sky from southern England (in the outskirts of London) and managed to discover and catalogue thousands of faint nebulae. Einasto and his Tartu Observatory colleagues needed to collect and measure the spectra of just 30 to 50 galaxies in the direction of a single carefully selected cosmic void to have challenged the priority of the discoveries Gregory and I submitted to the *Astrophysical Journal* in September 1977. Tartu Observatory researchers were suggesting the existence of giant "holes" in the galaxy distribution, but without substantiating evidence from galaxy redshift surveys, their claims were essentially a consistency argument because their early published papers had no definitive proof of these claims. The definitive proof that cosmic voids dominate the large scale structure in the galaxy distribution came when new redshift survey data was available from the West. This was the very basis of the advances made in this era, and the Estonians were not in a position to contribute.

6.2   Zeldovich Requests Assistance

The early work on the galaxy distribution at Tartu Observatory started in 1974, when Zeldovich approached Einasto (2014, pp. 122–3) and asked if Einasto could investigate the large-scale distribution of galaxies to distinguish between three theories of galaxy formation: Peebles' favorite hierarchical clustering, Ozernoi's cosmic turbulence model, and Zeldovich's favorite top-down pancaking theory. As described by Einasto, despite the fact that no one at their observatory had experience in observational cosmology in 1974, they got busy in learning what they might accomplish. With no access to new observations of their own, they compiled information from scientific publications and catalogues. This included known galaxy redshifts for peculiar and interacting galaxies (including what are called "Markarian galaxies") and for groups and clusters of galaxies (Einasto 2014, p. 124, p. 128). Their reaction to the challenge was appropriate, but it remains a matter of opinion to decide how much progress they made. The most generous way to describe their work is to say that the Tartu Observatory group began to recognize a skeleton structure that could be associated with the large-scale distribution of galaxies. One iron-clad statement can be made: when they found what appeared to be supercluster structure surrounding a cosmic void, they had no means to delineate the true structure. Proof is needed to distinguish empty regions from those regions that have simply been under-sampled (having not yet been surveyed). To prove that cosmic voids exist requires well-defined and complete redshift samples in the direction of the structure.

To investigate the 3D distributions of galaxies, the sample being studied must be well-defined and complete, and galaxy redshifts are essential to provide distance information. Once the completeness for a sufficiently faint sample of galaxy redshifts in a specific area of the sky begins to approach ~95%, all worries dissolve. At any level less than this, it is fair enough to ask whether an apparent "hole" in a 3D map results from an incomplete or imperfect data set. When the Arizona redshift surveys began, we also searched all available catalogues for galaxies with known redshifts; we could not afford to waste our telescope time remeasuring galaxy redshifts that were already known. What we found in the same mid-1970s time frame was that for structures located beyond our Local Supercluster, at most ~25% of the brighter galaxies in well-studied regions of the sky already had redshifts. The astronomers at Tartu Observatory were fooling themselves to think that a 3D analysis with less than 95% completeness was going to provide a convincing test as to whether the galaxy distribution was either homogeneous or filamentary and also to provide the basis to define its topology. A proper answer to these questions, especially in the presence of skeptical cosmologists, would require much higher levels of completeness than available to the Tartu Observatory astronomers in the period 1975 to 1977.

Einasto states in his monograph (Einasto 2014, p. 135, p. 142) that there were sub-samples of objects in the Tartu Observatory supercluster studies that were complete. In particular, he singles out the Zwicky Near Clusters to be one of these. (Near refers to being nearby.) Einasto suggests that the Zwicky Near Cluster sample formed a central part of their analysis. The late Mihkel Jõeveer (1937–2006), the first author on most of the early Tartu Observatory studies of superclusters, devised a method to estimate distances to the Zwicky Near Clusters based on the brightness distribution of galaxies that are assumed to be cluster members; astronomers call this the "galaxy luminosity function."

This was a fine idea, but there was one hidden problem. Zwicky galaxy clusters are a poorly defined sample and may be of little to no general use for well-defined 3D studies for the following reasons.

Zwicky and his assistants used a freehand qualitative method to define the outer borders of their clusters. These contours were based on what Zwicky saw in the apparent galaxy distribution on the National Geographic Palomar Observatory sky survey photographs on the day he was making his charts. This was described in Chapter 4 (Section 4.3). Shown in Figure 6.1 are two maps from a test of exactly the same nature. Here, I follow the lead of Vorontsov-Velyaminov (1987). The top map (Figure 6.1a) is from the early reference: Zwicky (1959). The bottom map (Figure 6.1b) is from the published Zwicky catalogue (Zwicky, Herzog, & Wild 1961). There is little to no resemblance between the cluster outlines that are drawn on these two maps despite the fact that they come from the same area of the sky (identically the same National Geographic Palomar Observatory Sky Survey photographic plate). This lack of resemblance in the Zwicky cluster identifications is consistent with Abell's separate test discussed in Chapter 4.

A careful quantitative check on the completeness of Zwicky clusters was made by Postman, Geller, and Huchra (1986), when they completed a two-point correlation analysis for clusters in Abell's and in Zwicky's cluster samples. One key result of the Postman et al. study was to show that Zwicky's Near Clusters are quite different from all other samples in that they "appear statistically equivalent to a random sampling of the galaxy distribution." In other words, Zwicky and his assistants did not do a good job defining the Near Cluster sample. Even if the Tartu Observatory astronomers did a perfect job studying this sample, the study is, by its very nature, ill-defined and incomplete. Of course, there will be galaxies in regions that Zwicky designated as Near Clusters, and there are likely to be no galaxies where there are no Zwicky Near Clusters, but Zwicky's method of identifying clusters was flawed, and the sample itself is not appropriate for studying the large-scale distribution of galaxies in a complete and definitive sense. There is simply no good substitute for brightness-limited complete redshift samples of individual galaxies to define and confirm the true 3D distribution of galaxies.

The nature of the nearby Zwicky clusters was put to yet another test by Dr. Harold Corwin, who worked for many years with de Vaucouleurs on the Reference Catalogues of Bright Galaxies and with Abell on extending the original Abell rich cluster catalogue into the southern hemisphere. In an email exchange with Dr. Corwin in 2019, I learned that Einasto approached Corwin in the early 1980s (in the very time that Corwin was working with Abell) and suggested that a counterpart to the Zwicky Near Cluster sample be compiled

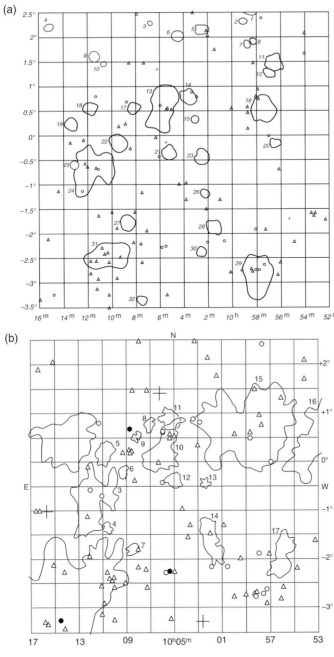

Figures 6.1a (top panel), 6.1b (bottom panel). Zwicky Near Clusters. These two maps represent a "blind" test of Zwicky's ability to draw the borders of his galaxy clusters. Each square plot represents a single plate from the National Geographic Palomar Observatory Sky Survey (described in Chapter 4) that covers an area on the sky of 6° x 6°. The meandering lines represent the outer borders of the galaxy clusters as identified by Zwicky, and there appears to be almost no correspondence between these lines in the two drawings. The Zwicky Near Clusters are, on average, the larger structures as seen primarily in the lower panel. Upper panel is reproduced from Zwicky (1959) with permission (Springer-Verlag OHG) and the lower panel is from Zwicky, Herzog, and Wild (1961).

from the southern sky survey photographs. Corwin tested this possibility by personally inspecting areas of the sky near the celestial equator where southern sky survey photographic plates happen to cover areas in the Zwicky cluster catalogue. This is what Corwin said to me about his attempt to cross-identify the newer sky survey plates with Zwicky's nearby cluster identifications: "It turned out to be a frustrating and ultimately futile exercise: the cluster cores were there but the isopleths (the outer cluster boundaries) matched Zwicky's about as well as randomly-drawn lines." When Corwin reported these results to Abell, Abell's comment was along the lines of "I told you so" and was dismissive of what he called Zwicky's "amoebas." Unfortunately, Einasto's suggestion went no further than this initial test of the concept.

Einasto also suggests that their Markarian galaxy subsample was complete and that it was useful in the Tartu Observatory studies of supercluster structure. It is somewhat of a coincidence that I happened to have studied the complete Markarian galaxy redshift sample as it relates to cosmic voids. My interest in these galaxies was piqued by a 1982 publication that claimed Markarian galaxies are uniformly distributed throughout the Universe and therefore that they may fill the voids (Balzano & Weedman 1982). To me this seemed implausible. My own analysis (Thompson 1983) shows that Markarian galaxies do not fill the cosmic voids. They appear to be somewhat uniformly distributed in space because they are rare objects: they undersample the galaxy distribution so you cannot see any supercluster structure with this sample. I was pleased to find a simple answer to the problem I investigated in 1983, but from the perspective of the Tartu Observatory group, this is bad news. Even with a complete Markarian sample, the supercluster structure cannot be detected with these galaxies because they undersample the galaxy distribution. In 1977, the primary focus of Einasto and Jõeveer was to study the galaxy distribution in the Southern Galactic Hemisphere in the vicinity of the Perseus-Pisces supercluster. My 1983 analysis shows that there are only 15 Markarian galaxies in and around the main body of the Perseus-Pisces supercluster, a supercluster that spans a length in the sky of more than 40 degrees and contains many hundreds of galaxies. Markarian galaxies may have helped somewhat with the Tartu Observatory analysis, but there is no way that this so-called complete sample adds confidence to the process of defining superclusters and cosmic voids. The fact that they are a complete sample is irrelevant.

At this point, the story splits into two different topics. One concerns the initial scientific paper that was published based on the Tartu Observatory supercluster work. The second concerns the Tartu results that were presented at IAU Symposium No. 79 in Tallinn, Estonia. These two topics are discussed separately.

6.3   Jõeveer, Einasto, and Tago: Tartu Observatory Early Effort

The Tartu Observatory scientists submitted their initial scientific results to the highly respected Western astronomy journal *Monthly Notices of the Royal Astronomical Society* on August 1, 1977 in a paper they entitled "Spatial Distribution of Galaxies and Clusters of Galaxies in the Southern Galactic Hemisphere" (Jõeveer, Einasto, and Tago 1978). Following a standard process that all scientists face, the editor of the journal forwards the manuscript to a referee he/she has selected who is familiar with the subject matter and has a reasonable sense of judgment. The referee, whose name is generally not revealed to the authors, reads the manuscript and reports his or her opinions and conclusions to the journal editor, who decides whether to accept, to accept with minor revisions, or to reject the manuscript. In the third circumstance, the authors read the referee's comments and have the option to resubmit a revised version of the paper. It is clear from the description in Einasto's 2014 monograph and from the title page of the published paper by Jõeveer, Einasto, and Tago (1978) that the Tartu Observatory's August 1, 1977 manuscript faced severe criticism and that it had to be significantly revised. They eventually submitted a revised manuscript on April 3, 1978. In both of his personal accounts, Einasto (2014 and 2018) describes his frustrations with the referee and discusses some of the revisions that were required.[2] However, by the time the revised version of their manuscript was submitted, IAU Symposium No. 79 was finished, and the three authors Jõeveer, Einasto, and Tago had heard at the symposium the details of the redshift surveys from the Arizona telescopes and our description of cosmic voids. Jõeveer et al. added a reference to our redshift survey work in their revised paper. By doing so, they reset the clock, from a priority perspective, and their published work cannot be considered independent of ours. They simply erased their claim of having first priority in terms of the discovery of cosmic voids in a refereed scientific journal. On the other hand, Gregory and I submitted our Coma/A1367 manuscript to the *Astrophysical Journal* on September 7, 1977. We did not attend IAU Symposium No. 79. We read no publications and no preprints from the Tartu Observatory group until after our paper was accepted for publication. We made only the most minor revisions as directed by the editor and the referee of our paper. Given our seven-month lead in submitting the published versions of the manuscripts (September 7, 1977 compared to April 3, 1978), it is no surprise that our paper appeared in print first.

6.4   Tallinn IAU Symposium, September 1977

Next, consider what happened at IAU Symposium No. 79 in Tallinn, Estonia, held September 12–16, 1977. Recall from Chapter 5 that Gregory (along with Tifft) submitted an application to attend the symposium. After receiving Gregory's request to attend, meeting organizer Malcolm Longair sent Tifft and Gregory a letter saying that Gregory would not be invited. Longair suggested that Tifft present and discuss our work at the meeting. Since Longair had never read our new redshift survey paper on the Coma/A1367 region, he certainly did not realize the significance of what he was doing. This had one minor and one major consequence. The minor consequence was that the Gregory and Thompson work, in several cases, was credited to Tifft and Gregory with reference to the IAU Symposium paper. In fact, Jõeveer, Einasto, and Tago (1978) made this mistake. Of greater importance is the point that neither Gregory nor I had a chance to provide our scientific input at the conference.

There was a long-standing tradition at some American and at European observatories to write "observatory circulars" or "observatory reports" before (or instead of) publishing research results in journals. From the 1950s to the 1970s, this practice was used at Tartu Observatory. A number of Einasto's and Jõeveer's results on superclusters appeared in this alternate "published" form. These circulars or observatory reports generally have no referees and no external control regarding dates of completion. At IAU Symposium No.79, Jõeveer, Einasto, and Tago circulated their preliminary results on superclusters in this format, with a 1977 date designation. Some have associated these 1977 "publications" with the discovery of cosmic voids, but this was the same material that would be heavily criticized by the referee appointed by the *Monthly Notices of the Royal Astronomical Society* and initially rejected for publication. Einasto (2014, 2018) mentions other minor publications (i.e., "circulars") dating back to 1975 that contain results of his work on superclusters. I emphasize that these are not certified in the style of Western scientific journals with a journal editor in a referee-based system.

Einasto (2014, pp. 138–9) describes the fact that preliminary copies of the Tartu Observatory supercluster studies were sent to Peebles at Princeton University, sometime before the IAU symposium. In response, Peebles sent Einasto a letter suggesting that the structures Einasto reported could be figments of the imagination where the eye connects random dots in a biased way. Peebles' response included two high-quality photos from Soneira and Peebles (1978), one showing the true Lick survey galaxy distribution from the northern galactic cap region and the second a carefully simulated computer-generated 2D distribution of "fake data" that had statistical characteristics of the Lick galaxy map (see Figures 6.2a and 6.2b). The immediate visual impression both photographs evoke is quite similar, and because Peebles could guarantee that the "fake data" map does not contain any true filamentary large-scale structure, he was in a position to suggest that the Universe (i.e., the Lick galaxy map itself) does not contain any such

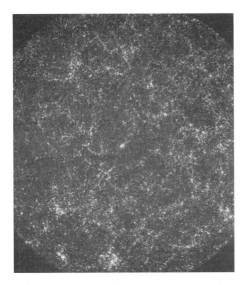

Figure 6.2a Lick galaxy map. This map shows the distribution of galaxies in the north galactic cap. The prominent white feature near the center is the merged images of objects in the Coma cluster of galaxies. It sits in the foreground, whereas nearly all galaxies in this plot are more distant. By permission of the A.A.S.: R. Soneira and J. Peebles (1978). *Astron. J.*, 83, pp. 845–60.

Figure 6.2b Lick galaxy map simulation. Soneira and Peebles' simulated galaxy distribution map. Additional discussion of this map is given in the text of both Chapters 6 and 7. By permission of the A.A.S.: R. Soneira and P. Peebles (1978). *Astron. J.*, vol. 83, 845 – 860.

structure, either. Peebles was the premier advocate of the statistically based hierarchical model of the galaxy distribution, so to him, this was a natural conclusion.

Forty years after these events transpired, with 20/20 hindsight, one can recognize that Peebles was making a false consistency argument. Of course, the real galaxy distribution does possess true structure in the form of filaments and sheets of galaxies, and yet the structure is indistinct in the Lick galaxy map (Figure 6.2a) because multiple structures are projected one over another, along the line of sight. The true structure becomes distinct when depth information is available from galaxy redshift surveys. It is important to add that when some contemporary astronomers and cosmologists saw the Lick galaxy map, they believed that the hints of filamentary structure were real, despite what Peebles was saying. One of these was Robert Dicke of Princeton University, Peebles' thesis advisor, who was interviewed by Allan Lightman for the Lightman and Brawer (1990, p. 209) book entitled *Origins*. Dicke said the following: "I was impressed by the appearance of those filaments in there. They seemed real. I kept arguing with Jim that they were real, and he kept saying that they were a figment of the imagination."

In the face of Peebles' criticism, however, Jõeveer and Einasto could not defend themselves. The only way to prove their case was to have further information, showing in 3D that cosmic voids are actually empty and that their so-called supercluster "cell" structure consists of bridges and sheets of galaxies. With insufficient proof of their own, their only recourse in their published manuscript was to reference redshift survey data from the Arizona group that was discussed and displayed at IAU Symposium 79. But to change the minds of traditional holdouts like Peebles would require much more extensive and decisive observational proof, and for eight long years – from 1978 to 1986 – Peebles stood firm.

With these preliminary facts clearly stated, it is unreasonable for Einasto to suggest – as he does in his 2014 monograph – that it was at the IAU Symposium in Tallinn that the discovery of cosmic voids was first established.

Mihkel Jõeveer (1937–2006)

Mihkel Jõeveer was an Estonian astronomer and research scientist who spent his entire professional career at Tartu Observatory. His early work concerned the distribution and dynamics of stars in our Milky Way galaxy, the topic of his PhD in 1984. He later became a team member with Jaan Einasto in the study of dark matter in individual galaxies and in the study of the large-scale distribution of galaxies. Biographical information is from Einasto (2014). Photo reproduced with permission: copyright Jaan Einasto.

It is very important to point out that part of the Estonian's "consistency argument" was based on the early Zeldovich pancaking concept as shown in Figure 6.3. As we will see in Chapter 7, the 1977 version of that theory was itself incomplete. It is absolutely true, however, that the general concept of cosmic voids in the galaxy distribution was openly discussed at the IAU Symposium in Tallinn. The concept of voids appears in the published scientific papers from

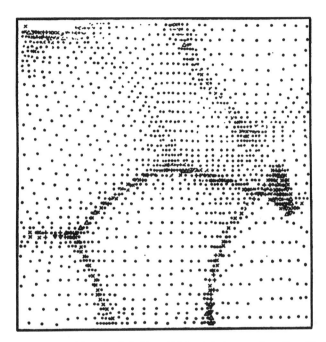

Figure 6.3 Shandarin's First Zeldovich Approximation Model 1975. This 2D simulation is the first to demonstrate that initial conditions defined by the Zeldovich Approximation (ZA) lead to extended supercluster-like structures. Ideally, each black dot would represent a galaxy and the linear features (collections of black dots), the filamentary supercluster structure. Although the model was incomplete – lacking dark matter, in particular – and does not accurately depict galaxy and supercluster formation exactly as we know it today, Jõeveer and Einasto used this diagram as the basis to search for supercluster structure. The diagram is highly significant theoretically, as it is the first simulation to incorporate the ZA. Reproduced by permission of Oxford University Press: Einasto, J., Jõeveer, M. and Saar, E. (1980). *Mon. Not. Royal Astron. Soc.*, 193, pp. 353–75.

that meeting (all of which were sent to the conference organizers *after the meeting*, following standard procedures). Recall that before this meeting, we had requested Tifft to introduce and openly discuss our Coma/A1367 redshift survey at the meeting and encouraged him to use the word "void" to describe this phenomenon. All in attendance were aware of what had been discovered, but some – including meeting organizer Malcolm Longair – made the mistake of concluding that the discovery was made at the meeting, not recognizing that the Gregory and Thompson Coma/A1367 manuscript arrived at the *Astrophysical Journal* before this meeting began. The meeting participants were discussing results that came directly from data that Gregory and I had already published. Gregory and I requested Tifft to discuss this new phenomenon, and one outcome of the meeting was for Einasto to make the most of Gregory's and my results as well as our absence.

Another point worth mentioning is spelled out in Einasto (2018, p. 24), where he states " . . . most observers studying the distribution of galaxies concentrated their efforts on examining the environment of rich clusters (Chincarini & Rood 1976, Gregory & Thompson 1978)." He goes on to say that wide field surveys are needed to understand the cosmic web that cover a "full range of galaxy densities from the richest clusters to the emptiest voids" at which point he praises the second CfA redshift survey (de Lapparent et al. 1986). Those of us from Arizona knew this, too, but there is a practical side to scientific work. What is needed and what is immediately feasible do not necessarily overlap: we had limited access to observing time on large telescopes. Furthermore, when our early redshift surveys began, it was the distribution of galaxies in the 100 Mpc *depth* of our first redshift survey (along the line of sight to the rich clusters) that revealed "the richest clusters and the emptiest voids." This was the hidden beauty contained in the Gregory and Thompson observing proposal displayed in Appendix A. Before the observations began, we realized, based on the 1976 precursor work of Tifft and Gregory (1976), that some of the more interesting results from our proposed Coma/A1367 redshift survey would come from the galaxy distribution along the line of sight in the foreground volume. That aspect of our work paid off handsomely. Furthermore, as summarized in Chapter 8, the sum total of our redshift survey work – all of which preceded the de Lapparent CfA Slice of the Universe redshift survey – was closely equivalent to what Geller and her collaborators would publish in their first CfA2 slice. We investigated long swaths of sky in both the northern and southern galactic hemispheres. We called the northern hemisphere swath the "Hercules and A2197+A2199 Broad View" (shown in Figure 8.1). It is a 46°-long survey that cuts across a large volume of space, a majority of which shows a low density of galaxies. The redshift survey plot in Figure 8.1 was published in Gregory and Thompson

(1984). So Einasto is incorrect to say that our early work (work that came before de Lapparent et al. 1986) was concentrated on simply examining the environment of rich clusters. We investigated a full range of environments, even while working within the constraints of limited telescope time.

Two balanced accounts of the events surrounding the cosmic void discovery were discussed in Chapter 5. One is by Oort (1983) and the other is by Sandage (1987). Oort's review article on "Superclusters" includes a discussion of the published results from the Jõeveer, Einasto, and Tago (1978) paper showing their framework (or skeleton) of the Perseus supercluster, but Oort does so in a section that includes both the complete brightness-limited redshift survey data of Gregory, Thompson, and Tifft (1981) as well as preliminary 21-cm radio wavelength observations of Giovanelli, Haynes, and Chincarini (1986). In the description by Sandage, he does not highlight the work from Tartu Observatory, even though I know he was fully aware of it. The Oort (1983) review was written long after the 1977 IAU Symposium Meeting in Tallinn, Estonia. Even in 1983, Oort does not accept exaggerated claims like those in Einasto's 1977 conference presentation that "the Universe has a cell structure." Here are Oort's words from 1983: "It has even been hypothesized that they (the superclusters) may all be interconnected, so that the Universe would consist of a three-dimensional network of superclusters, with essentially empty meshes in between. Presently available data are, however, quite insufficient to trace such a network." Notice that Oort does not argue against the interconnected 3D network, but he cautiously states (in 1983) that the available data are "insufficient to trace such a network." This statement reveals the basic fallacy in the early Tartu Observatory work. They had made an educated guess – based on the profoundly important theoretical work shown in Figure 6.3 by Zeldovich and Shandarin – that the observations of the galaxy distribution would show a cell-like structure, but the observations were inadequate to confirm it in 1977 and still marginally so in 1983. Our work from the Arizona telescopes had shown empty voids and filaments of galaxies stretching across vast distances, but the hypothesized cell structure was still something for the future. Until massive dedicated redshift surveys began like those at the Center for Astrophysics, the redshift data samples were insufficient to trace such a network in the way Einasto envisioned. In 1977, Einasto was making a consistency argument regarding voids that eventually would be proven true, but the topology he suggested as his best guess was wrong.

In summary, neither Oort nor Sandage put significant weight on IAU Symposium No. 79 in terms of its role in the discovery of cosmic voids or in the effort to define the large scale structure in the galaxy distribution. Even so, Einasto keeps returning to his story over and over again that the events at the meeting were especially significant. Scientific meetings do have a role to play in the scientific process: they provide a forum for the exchange of either published or interim research ideas among those who attend the meeting. However, scientific meetings are not the place where astronomical discoveries are made.

6.5   Brent Tully in Tallinn and the Local Void

Two years after the 1977 Tallinn IAU Symposium, I accepted a position at the University of Hawaii's Institute for Astronomy. I arrived in Honolulu the last weekend of July 1979 and showed up at the Institute for Astronomy on August 1. One of the first astronomers I met was Brent Tully (b. 1943). He was excited to show me the Local Supercluster movie he had created two years earlier for the Tallinn meeting. The day I arrived, Tully set up a computer with its display terminal in the second-floor foyer near his office, and I watched his movie with several other astronomers and graduate students looking on. I discuss it here because in Einasto's monograph, he cites Tully's movie as another example of results presented at Tallinn in support of the new vision of the inhomogeneous local galaxy distribution (Tully & Fisher 1978a, 1978b). Sure enough, the Tully video was an interesting eye-opener. By this time, Tully had spent a decade determining distances to galaxies in our local neighborhood of the Universe, cataloguing them, and analyzing the resulting data set. In this regard, it became clear that he was following in the footsteps of Gerard de Vaucouleurs (e.g., Tully 1982). Tully's 1977 video featured the 3D distribution of galaxies in the Local Supercluster, and it was special because the computer movie rotated that distribution around an axis perpendicular to the flat plane of the Local Supercluster, so it appeared to the viewer that the supercluster was rotating at a rate of once every ~30 seconds in the computer display. The supercluster plane looked crisp and prominent as it appeared to rotate, and there were several distended diaphanous filaments of galaxies (they looked like bloated symmetrically shaped fingers) that seemed to point toward the Virgo cluster itself situated at the center of the supercluster. But these diaphanous structures bore no relationship to the flattened part where we reside in the Milky Way galaxy.

It was not clear to me, when I saw the video clip in August 1979, exactly how the impressive planar nature of the Local Supercluster was related to cosmic voids and to the other features in the large scale distribution of galaxies that Gregory and I had discovered just a few years earlier. The general discussion of cosmic voids at IAU Symposium No.79 in Tallinn put Tully in a perfect position to first conceptualize, and then to begin discussing the Local Void and the supercluster plane in this broader context.[3] It would be twenty-five to thirty years later – after a great deal of further work on the part of Tully and his collaborators – that the flat planar galaxy distribution in which we live would be explained and then beautifully integrated into the grander scheme of the large-scale galaxy distribution. The details of this final step are left for Chapter 9.

R. Brent Tully (b. 1943)   Brent Tully comes from Canada. He received his PhD degree at the University of Maryland for completing an analysis of the internal motions of gas in the galaxy M51 (the galaxy shown Figure3.4) measured at optical wavelengths. Subsequently, he and fellow Maryland graduate student J. Richard Fisher used the facilities at the National Radio Astronomy Observatory (NRAO) to complete the first comprehensive 21-cm neutral hydrogen survey of galaxies in a volume of space that encompasses the entire Local Supercluster (extending to a Doppler velocity of 3,000 km s^{-1}). This provided a rich source of data that Tully has incorporated into many creative research projects. Tully spent two years at the Observatoire de Marseille before he accepted a research astronomer position at the University of Hawaii Institute for Astronomy, where he has remained for the remainder of his career. He is the codiscoverer of the fundamental Tully-Fisher Relation, which relates the measured 21-cm velocity-widths of late-type galaxies to their total intrinsic luminosity. Tully used the NRAO 21-cm Doppler velocity observations along with optical wavelength observations to create the "Catalogue and Atlas of Nearby Galaxies," the most detailed census to date of galaxies in the Local Supercluster. In the past decade he has worked with a number of collaborators to catalogue distances to 18,000 galaxies in the local Universe and to use them to study, with his current collaborators, velocity flows of galaxies in and around the Local Supercluster (Shaya, Tully, Hoffman, & Pomarede 2017) and beyond (Pomarede, Hoffman, Courtois, & Tully 2017; Tully et al. 2019). Photograph reproduced with permission: copyright Igor Karachentsev.

7

Theoretical Models of Galaxy Formation – East versus West

Throughout the 1970s and into the 1980s, two primary schools of galaxy formation were being pursued: one was called "bottom-up" and the other "top-down." Both were introduced and briefly discussed in Chapter 2. Year upon year, each school tested their model predictions against observations of galaxies. For the bottom-up advocates, the tests were primarily statistical. For example, they asked what is the chance of finding another galaxy in the near vicinity of a specific set of target galaxies (i.e., the galaxy correlation function), or what is the velocity distribution among random "field" galaxies? Answers to these questions placed restrictions on the best model. The top-down school often used evidence from the large-scale distribution of galaxies like the general appearance of cosmic voids and filamentary structure in the galaxy distribution. The competition between these two models became a point of interest both to those who doubted the reality of filamentary structure as well as to those of us who first identified it.

The story begins in 1970 at a time when dark matter played no role in galaxy formation. The traditional Western model, based on bottom-up evolution, was being studied and developed in Princeton, New Jersey, and a newer upstart model based on top-down processes emerged from work done in Moscow. At Princeton University, P. J. E. Peebles was the key proponent, and in Moscow it was Yakov Zeldovich. Both of these scientists trained and then worked with excellent students. As the research effort moved forward, others joined the endeavor and contributions were also made from institutions including the University of Texas, Tartu Observatory, Oxford University, and the University of California, Berkeley. In the end, cosmic voids played a key role in the selection of a hierarchical-based CDM model as the final compromise solution and the model that is most favored today.

The cosmological setting in the 1970s was as follows. Astronomers acknowledged there was "missing mass" associated with rich galaxy clusters, but few anticipated the profound significance and influence dark matter would have as a central driver of galaxy formation. Based on the measured abundance of light elements like deuterium and lithium, in the mid-1970s astronomers began to realize that baryons (the ordinary material in our bodies and in the stars) may not provide enough mass by themselves to make the geometry of the Universe closed. Therefore, in the 1970s, some astronomers accepted a low-density Universe with a geometry that was "open" while others assumed a "closed" geometry, even though evidence for a low baryon density ran contrary to that assumption. Studies of the Cosmic Microwave Background (CMB) were sufficiently primitive in 1970 that the point-to-point irregularities (as detected on the sky) of the CMB radiation remained undetected. Even in this early era cosmologists recognized that the amplitude of the CMB irregularities (its apparent bumpiness or texture) could provide hints as to how galaxies formed, but only upper limits were available as this story begins.

The Princeton school promoted the bottom-up model of galaxy formation in which the earliest objects to form have low mass. Low mass in this context means clusters of stars with total masses in the range of ~1/2 million times the mass of our Sun. The Princeton concept was labelled bottom-up because these low-mass star clusters were hypothesized to be the starting point in a sequence of coalescence events where larger and larger objects are built by the natural merging of smaller ones. Eventually, galaxies and clusters of galaxies are produced in the merger process. On the other hand, the Moscow school promoted a top-down model in which the earliest objects to form were massive gaseous "pancakes" that were the size of entire superclusters. As these pancakes form, they would fragment into a range of smaller galaxy-sized units. The minimum mass of an entire supercluster is ~5×10^{14} times the mass of our Sun and is, therefore, 100 million times larger than globular clusters. The huge pancakes formed sufficiently far back in the past that their initial gravitational collapse and fragmentation would not have been seen directly with a telescope.[1]

Even though these two schools are now part of history, the face-off between them became intertwined with the discovery (and the acceptance) of cosmic voids and supercluster structure. The challenge – for those cosmologists who accepted the reality of the void and supercluster observations – was to figure out how this structure meshed with the best galaxy formation model. At the earliest stages, the supporters of the hierarchical model simply ignored the structural features in the galaxy distribution, but when galaxy redshift surveys revealed more and more detailed structure, it became a primary focus of attention for testing the models.

At first, the top-down model seemed to have the upper hand in explaining the void and supercluster structure because this model suggested how primeval material might have been organized over huge supercluster scales from the earliest times. Ironically, although the original predictions of the top-down model looked somewhat similar to the observed configuration of voids and superclusters, the details of the model were far from being correct. Simultaneously, the proponents of the bottom-up model impeded progress by suggesting that it was unlikely that the observed large-scale structure in the galaxy distribution was real. When extended filamentary features appeared in galaxy maps, Peebles blamed it on the tendency of the eye to artificially create linear features that (in his view) did not exist. No doubt he thought the situation was somewhat like the "canals" on Mars that had been reported by some astronomers from the 1890s through the 1930s. In the explanation that follows, the bare essentials of both theories of galaxy formation are described, and I explain the influence each model had on the development of our understanding of cosmic voids and the large-scale structure.

7.1   Bottom-Up Theory a.k.a. Hierarchical Clustering

The bottom-up theory advocated by Peebles was based on traditional ideas. Near the end of Chapter 3, I summarized the accepted view of galaxy and cluster formation circa 1937. At that time, the Swedish astronomer Erik Holmberg published his observations of double galaxies, and to explain what he saw, he hypothesized that when two galaxies pass close to each other, they interact gravitationally via tidal forces and can, potentially, become a captured pair resulting in one galaxy orbiting the other. Holmberg (and others in that era) suggested that pairs of binary galaxies would eventually meet other pairs of galaxies producing a group of four, and as time passed, larger groups and clusters of galaxies would form.[2] This concept was the basis of hierarchical galaxy formation. But in the 1930s no one had the slightest idea how the process began. The Princeton group built the foundation for this theory and fleshed it out.

Peebles has stated that his primary attraction to the hierarchal clustering model was the simple one-to-one relation between the formation times of galaxies in the hierarchical model and the ages of the oldest stars astronomers have discovered. In 1968, Peebles published a paper with Robert Dicke (1919–87), Peebles's PhD thesis advisor, suggesting that globular star clusters (see Figure 7.1) might be the first major systems of stars to form in the early Universe (Peebles and Dicke 1968). If so, globular star clusters could provide the initial building blocks for bottom-up galaxy growth. This idea made good sense fifty years ago when evolutionary

Figure 7.1 Globular Cluster M15. This beautiful object is one of ~150 globular
star clusters that orbit within the halo and the central regions of our Milky Way
galaxy. Peebles and Dicke hypothesized in 1968 that star clusters like M15 were the
first objects to form in the Universe and that globular star clusters could provide the
basic building blocks for hierarchical galaxy growth. The more massive globular
clusters like M15 contain approximately 1/2 million stars. The cluster is held in
a nearly spherical configuration by the self-gravity of the entire star cluster system.
Twenty-five-minute exposure taken by the author, September 1981, with the prime
focus camera of the Canada. France Hawaii Telescope, Mauna Kea Observatory,
Hawaii. With permission: copyright Canada-France-Hawaii Telescope Corporation.

models of stars had just shown that globular clusters contain stars that are
among the oldest known.

The increasing power of digital computers in the 1970s created new oppor-
tunities to demonstrate hierarchical growth of galaxy clusters with N-body
simulations. As we will see, Sverre Aarseth of Cambridge University was one
of the major pioneers. Another was Peebles himself. Among the significant early
papers by Peebles was his simulation of galaxy cluster formation (Peebles 1970).
In his simulation of the Coma cluster of galaxies 1,000 (identical) point masses –
each one representing a single galaxy – were set free to move in 3D under their
mutual force of gravity. Just as predicted by the hierarchical model, the point
masses in the model eventually organized themselves into what appears to be
a giant galaxy cluster. Another very significant result came in 1974 when
William Press (b. 1948) and Paul Schechter determined how objects formed by
hierarchical clustering would be distributed in relation to their mass, that is, the
number of small objects relative to the number of massive ones (Press and
Schechter 1974). In 1976, Simon White went a step further and extended the
early work of Peebles on cluster formation by following 700 test particles that
were given a range of masses (White 1976). White noticed, for the first time,

what he termed "the continual formation and amalgamation of sub-condensations" among the particles (i.e., the galaxies) in his model. Three years later, another even more ambitious simulation, also supporting the hierarchical model, was built by Aarseth, Gott, and Turner (1979). By this time, Aarseth had spent many years tuning his computer skills to obtain the optimal scientific results from what was, compared to today, very limited computing power. The Aarseth calculations involved tracking the paths of between 1,000 and 4,000 particles, each one representing a single galaxy. The Aarseth, Gott, and Turner results must have pleased Peebles because this study demonstrated the growth of groups and clusters of galaxies with no indication of any filamentary structure. Even so, Gregory and I were pleased with the 1979 Aarseth et al. results because their models showed evidence of cosmic voids, a phenomenon that Aarseth and his collaborators clearly described and compared to the cosmic voids Gregory and I had discovered. Why the model of Aarseth and his two collaborators failed to show filamentary and pancake-like structures in the galaxy distribution is a subtle but very significant point. Their starting conditions (the initial positions and velocities of their test masses) were not consistent with what we know today. If a proper set of starting conditions had been used based on the theoretical ideas of Zeldovich as fully implemented by Shandarin (see Figure 6.3), filamentary structures would most likely have been seen.

Peebles made additional contributions to the hierarchical theory in two other areas. In the first, with graduate student Jer Yu, he analyzed theoretically, how random disturbances (pressure waves) evolve in the early Universe and predicted at what level the cosmic microwave background might show point-to-point irregularities when measured on the sky. Over a one arcminute span (about 1/30 the diameter of our Moon), Peebles and Yu (1970) predicted fluctuations in the detected "radio wave antenna temperature" of at least $\delta T/T \sim 0.00015$. (Radio telescope measurements are often discussed in terms of the "antenna temperature," but it is essentially a measure of the power received at the telescope's focal plane.) This set a clear goal for astronomers who worked to assist cosmologists to refine measurements of the CMB with radio telescopes. Observations of the sky were beginning to reach this level of precision.

Second of all, with the assistance of his Princeton graduate students, Peebles analyzed statistically, the distribution on the sky of both galaxies and clusters of galaxies. Their primary statistical measure was the "correlation function." This all-important function defines the chance of finding a second object within a specified distance (most often, an angular distance on the sky) of a pre-defined reference object. For example, Peebles asked if Abell rich clusters are statistically correlated with one another. If they are, superclusters are real objects. If they are not, superclusters do not exist. In an effort to characterize the galaxy distribution in terms of the correlation function, Peebles and his students resurrected several published catalogues of galaxies, the most important of which was the Shane and Wirtanen survey from the early 1950s (see Chapter 4). Even though the Shane galaxy counts were originally presented and analyzed on a one-square-degree grid, Peebles acquired from Shane the original counts that had been made on a 10-arcminute-square grid. By deriving the galaxy correlation function for the extensive Lick survey galaxy sample and finding it consistent with his own ideas of hierarchical clustering, Peebles concluded that the galaxy distribution was exclusively in the realm of statistically random processes.

P. J. E. Peebles (b. 1935)

Jim Peebles was both born and received his early education in neighborhoods near the center of Winnipeg, Manitoba. He completed his BS Degree in Physics at the University of Manitoba, in 1958. From there, he went directly to graduate studies in Physics at Princeton University where he received his PhD in 1962, under the direction of Robert H. Dicke. His work with Dicke led to the rapid acceptance of the Penzias and Wilson (1965) millimeter-wave discovery of the cosmic microwave background radiation. As Peebles himself points out, Penzias and Wilson are to be credited primarily for the exhaustive search for alternative sources of noise in their microwave system (noise that was seen starting in 1959). Professor Peebles started as a Princeton Assistant Professor in 1965 and was quickly promoted to Associate (1968) and then to Full Professor (1972). He is now the Albert Einstein Professor Emeritus at Princeton University. In addition to being a prolific author of research papers in astrophysics, he has written three highly influential monographs used in the study of the Universe: *"Physical Cosmology"* (1971), *"The Large Scale Structure of the Universe"* (1980), and *"Principles of Physical Cosmology"* (1993). His newest book is entitled *Cosmology's Century: An Inside History of Our Understanding of the Universe* (2020). Because of his excellent work in astrophysics, Peebles has received the highest accolades any scientist could attain. Among his more successful PhD students are Jer Tsang Yu (City University of Hong Kong, retired), Stuart Shapiro (University of Illinois), and Margaret Geller (Harvard University Center for Astrophysics). Primary source: Peebles (1984). Image reproduced by permission of Princeton University: copyright Richard W. Soden photographer.

By the late 1970s, Peebles appears to have become so confident with the success of his own work that he began to suggest to those observers who identified filaments or extended sheets of galaxies that they suffered from a "tendency of the eye to see patterns in the noise." As best I can tell, this statement first appeared in a 1978 publication by Soneira and Peebles (1978) where these two authors created a model of a 3D galaxy distribution projected onto a 2D sky map that showed the simulated galaxy distribution. By carefully selecting a small number of adjustable parameters to match the observed correlation function of galaxies from the Lick survey itself, they managed to make their model closely match the appearance of the true galaxy map (both displayed a similar "frothy" appearance). Images of the real and simulated maps are described and displayed in Figure 6.2. With these results in hand, Peebles became strident in criticizing those of us who were using redshift survey data to map the local galaxy distribution. To use his own words, Peebles (1988) became vitriolic. Ironically, he was tricking himself in the sense that he perceived random processes everywhere, even when there was true structure to be discovered.

But significant obstacles began to arise. Chief among them were measurements of how smooth the cosmic background radiation appears on the sky. If the radiation is too smooth – it turns out to be very smooth – then the initial irregularities needed to trigger simple hierarchical growth in a matter-dominated cosmological model are not large enough to explain the structure we see in the Universe today. As mentioned above, Peebles and Yu had set the limit on the measured antenna temperature at microwave wavelengths at $\delta T/T$ ~0.00015, and the CMB measurements were showing numbers smoother than this limit. At the same time, our Arizona galaxy redshift survey observations – as described in Chapter 5 – repeatedly showed features in the 3D galaxy distribution that were difficult to explain with the conventional hierarchical model.

7.2   Nature of the Initial Irregularities

The two theories of galaxy formation, Peebles's and Zeldovich's, each assumed a different form for the initial irregularities that were imprinted in the distribution of the matter and radiation at the earliest times. These irregularities – in some sense they resemble random sound waves – are necessary because, if none were present, galaxies and other structures simply would not form. Peebles hypothesized that isothermal irregularities were responsible for the formation of the first objects, including globular clusters. Isothermal irregularities mean that the background photons (the high-energy gamma rays from the hot early Universe) remain undisturbed all the while the commingled matter (consisting of protons, neutrons, electrons, and neutrinos) contain small and random irregularities throughout the volume of the Universe. Proponents of this model pointed out in the 1970s that the primordial radiation field (the gamma rays) was so dominant that small built-in intrinsic matter irregularities had few consequences at early times, but as the dominance of the radiation field decreased due to the expansion of the Universe, matter irregularities would emerge. The term "isothermal" refers to the absence of a connection between the hypothesized irregularities in the matter and irregularities in the gamma rays. In other words, the radiation maintains a single uniform temperature (it is isothermal) while the matter contains the intrinsic perturbations. Once present, isothermal perturbations do not easily dissipate, so even small objects like globular clusters can eventually form.

To explain the formation of larger objects like galaxies and galaxy clusters, Peebles suggested "adiabatic" irregularities, and the Moscow school favored "adiabatic" irregularities entirely. In this case, the background photons and the commingled matter together are both disturbed by slight pressure waves that fill the Universe. Soon Joseph Silk (b. 1942) recognized that adiabatic perturbations dissipate over time when the squeezed gamma rays (those in the denser pressure disturbances) slowly drift out of their perturbations and drag with them charged particles (electrons and protons) in a process that is now called "Silk damping." This has the dramatic effect of smoothing the smallest adiabatic perturbations more quickly and leaves in place, only pressure waves that trigger the formation of the largest objects. Silk damping is a basic feature of the original Zeldovich top-down galaxy formation model because pressure waves that lead to supercluster formation (the largest irregularities) remain intact while the smaller ones that might otherwise lead to early individual galaxy formation are smoothed away. Peebles accepted Silk damping, too, but he invoked isothermal irregularities for smaller objects like globular clusters, and therefore, star clusters were still able to form on the smallest scales.

Notice that the descriptions given so far for the origin of the initial irregularities totally omit any mention of dark matter. Further discussion near the end of this chapter will clarify how the initial irregularities need to be redefined, once dark matter is considered as a key factor. Dark matter irregularities have the admirable feature of beginning to grow at an early epoch even when the matter and radiation continue to remain smooth. This provides a head start for forming clumps of dark matter (and, therefore, galaxy formation) and yet circumvents the objections that the observed $\delta T/T$ in the background CMB radiation appears to be too small to trigger galaxy formation.

7.3   Top-Down Theory a.k.a. Zeldovich Pancakes

Zeldovich (1970) published, in the Western journal *Astronomy & Astrophysics*, a set of very simple mathematical equations that describe the slow 3D "collapse" of mass in the early Universe. His new contribution involved the concept that a shrinking cloud of collisionless particles would become smaller at different rates in the three different axes x, y, and z. The textbook view of gravitational collapse prior to 1960 was one in which the shrinking object is reduced equally in all directions simultaneously, like a shrinking balloon. Zeldovich pointed out that if each of the three axes has its own separate collapse rate, different shapes will be created. Say, the collapse rate in the x-axis is the highest. The resulting object will shrink slower in the other two axes and become a thin sheet, thereby creating a pancake-shaped form. If two axes are equally rapid in their collapse and one is significantly slower, the resulting object will be prolate (a cigar or filament).

Several new lines of investigation eventually appeared based on Zeldovich's initial concept. One of the earliest and best known of these was a study that appeared in 1972, coauthored with one of Zeldovich's star students, Rashid Sunyaev (b. 1943). Their paper was entitled "Formation of Clusters of Galaxies; Protocluster Fragmentation and Intergalactic Gas Heating." Sunyaev and Zeldovich (1972) calculated what would happen if a huge cloud of ordinary matter (i.e., baryons) collapsed into a thin pancake-like sheet with dimensions and mass equal to that of an entire supercluster of galaxies. They aimed to predict the fate of the gas that falls into the "pancake" and found that the gas would settle into separate layers: a cool inner layer along the mid-plane (with a temperature somewhat less than 10,000 degrees K) and two outer hot, low-density shock-heated layers that would radiate at soft X-ray wavelengths. It was the innermost layer that Sunyaev and Zeldovich suggested would fragment into individual galaxies. Details of the fragmentation and the galaxy formation process were worked out in other publications (e.g., Doroshkevich, Shandarin, and Saar 1978). While their calculations were done properly, the model was wrong in the sense that ordinary matter in the early Universe never undergoes the processes discussed in their paper. Instead, the dark matter (which was not considered in the Sunyaev and Zeldovich model) is first to collapse. But their model was a reasonable starting point to introduce alternate galaxy formation scenarios.

Yakov Zeldovich (1914–87)

Russian theoretical physicist Yakov Zeldovich was born in Minsk, Belarus, and as an infant moved with his family to St. Petersburg. He was self-educated, never receiving an undergraduate degree, but he did attend postgraduate courses at St. Petersburg State University. As a young man he worked at the Institute of Chemical Physics of the USSR Academy of Sciences and, among many other things, became an expert in chemical deflagration. In 1963, Zeldovich moved from Sarov, a closed city where the Soviet Union's nuclear weapons were developed, to Moscow, where he became head of a division in the Institute of Applied Mathematics of the USSR Academy of Sciences. At that time, he began to investigate problems in astrophysics and cosmology and also began to recruit an elite group of talented young mathematical physicists who worked, learned, and received their PhD degrees under his supervision. One of his most successful early students was Doroshkevich, who became a permanent member of his group and a long-term contributor to theories of galaxy formation. Because Zeldovich belonged to the elite group of Soviet scientists who designed and developed the first nuclear bombs in Russia, he was held in the highest regard in authoritarian-based Soviet society. He was truly an outstanding theoretical physicist with multiple scientific interests ranging from the physics of black holes to the origin of galaxy superclusters. Zeldovich was so prolific that the late Stephen W. Hawking once said to Zeldovich: "Before I met you, I believed you to be a 'collective author' ...." Reference and photo credit: Ginzburg (1994).

I was at the critical stage of defining my PhD thesis research when I first saw the Sunyaev and Zeldovich paper in the journal *Astronomy & Astrophysics* in 1972. There it was right on the page: an alternative to the traditional hierarchical galaxy formation model and one that involved cluster-scale and even supercluster-scale organization of matter. I aimed to search for intrinsic alignment of galaxy rotation axes over dimensions as large as galaxy clusters. While my completed PhD thesis reported some effects of this nature, a more significant intrinsic alignment result came from my analysis of the Perseus supercluster redshift survey data as reported in Gregory, Thompson, and Tifft (1981). In the Perseus supercluster, there is a clear sign of supercluster-scale galaxy alignment (relative to the overall orientation of the supercluster structure) that Oort (1983), in his review of superclusters, cited as support for the Zeldovich pancake theory. Even so, as I said above, the Sunyaev and Zeldovich theory of galaxy formation is of historical interest only. Dark matter was neglected, even though we know today it is the dominant factor in galaxy and supercluster formation.

For several years in the early 1970s Vincent Icke of Leiden Observatory, the Netherlands, independently suggested (Icke 1973) a very similar model in which superclusters form first and galaxies subsequently condense or fragment from within the supercluster. Icke's work references theoretical predictions by Western scientists (Lynden-Bell 1963, Lin, Mestel, and Shu 1965) who had predicted, before Zeldovich, the outcome of nonspherical gravitational collapse of gaseous objects. Although the work by Icke was published in 1973, he appears to have been unaware of the Zeldovich 1970 contribution. Instead of invoking a shock–wave-induced fragmentation of the massive collapsing supercluster to form individual galaxies, as Sunyaev & Zeldovich had done, Icke suggested that turbulent motion might initiate the fragmentation process. Although neither Icke nor Zeldovich referenced each other's work, I recognized and discussed both models in my PhD thesis in 1974 that featured the origin of galaxy angular momentum.

While Icke put supercluster formation work aside for about 10 years, the Russian theorists moved forward rapidly. A new avenue of investigation was opened in Moscow by Doroshkevich and his research team that now included the young Sergei Shandarin (b. 1947), another excellent student who joined the Zeldovich research effort in the early 1970s after graduating from the Moscow Institute of Physics and Technology. The Doroshekevich group aimed to model the dynamics of the gaseous material in the early Universe – still ignoring dark matter – over a volume that included several "pancakes." This effort can be compared to and contrasted with the 1979 work on hierarchical structure formation by Aarseth, Gott, and Turner described earlier. Both were being done at about the same time. As a first step the Moscow group worked out mathematically the equations of motion that define the paths of their test particles under the all-important "Zeldovich approximation" thus giving their simulation a very realistic starting point. Then the Doroshkevich group built computer models to follow the dynamics of their test particles to show, in the period 1975–78, how matter responds to the initial adiabatic irregularities associated with pancake formation. Because computer capabilities were limited, the first models in Moscow were done in 2D and the graphics were printed on a line printer. The test particles responded to all surrounding masses and accelerated (or decelerated) accordingly as the matter gravitationally collapsed into pancakes. The basis of their computer calculations was – unlike that of Aarseth, Gott, and Turner – a technique that had been developed by Hockney, called the "Cloud-In-Cell" or CIC method. The first 2D model by Doroshkevich and Shandarin had 64×64 cells and 4,096 test particles. These test particles experienced the initial collapse and moved on, unlike the Sunyaev and Zeldovich calculation where gaseous shock fronts were created at the time of collapse.

Sergei Shandarin (b. 1947)

Sergei Shandarin was born in a small village near Moscow to a working-class family. At age seven he moved with his parents to Moscow, and at age fifteen he was admitted to a school specialized in mathematics and physics. This led him to Moscow Institute of Physics and Technology (MIPT) – one of the best universities in the country. As a fourth-year student of MIPT, he began regularly attending the seminars in astrophysics and cosmology at the Institute of Applied Mathematics (IAM) and at Sternberg Astronomical Institute both led by Ya. B. Zeldovich. Shandarin became one of his pupils and after obtaining a PhD degree in physics, joined the Zeldovich group at the IAM. In January 1987, he moved to the Theoretical Department of the Institute of Physical Problems, where Zeldovich had become the chair a few years earlier. After the death of Zeldovich, Shandarin moved to the University of Kansas in 1989. Shandarin's research stemmed from the revolutionary Zeldovich approximation (ZA) that obtains at the beginning of the nonlinear stage of gravitational instability. The numerical model based on the ZA developed by Shandarin in 1976 revealed the major geometrical and topological features of the cosmic web: the thin web-like concentrations of mass and vast regions of very low density between walls. This was followed by work with V.I. Arnold that provided a strong mathematical basis of the ZA predictions. The first N-body simulations conducted by Shandarin with coauthors in 1980–83 demonstrated the universality and stability of the web structures with respect to the shape of the power spectrum of the initial density perturbations. Shandarin with coauthors pioneered the application of catastrophe theory, percolation statistics, the adhesion approximation based on Burgers' equation, and partial Minkowski functionals in the theory of the cosmic web and the analysis of the corresponding observational data. In 2012, he and his coauthors suggested a radically new method of conducting and analyzing cosmological N-body simulations based on phase-space tessellations. He is a fellow of the American Physical Society. Photo reproduced with permission: copyright Sergei Shandarin.

According to Einasto (2014, p. 127), preliminary results from these 2D calculations (see Figure 6.3) were made available by Shandarin to the Tartu Observatory astronomers as early as 1975, and a version of these results was inserted into a review paper written in Russian by Zeldovich, Doroshkevich, and

Sunyaev (1976) with credit to Shandarin. A description of their mathematical method was first published in a Russian journal (Doroshkevich and Shandarin 1978) with a follow-up paper in a British journal (Doroshkevich et al. 1980). Their results suggest a cell-like structure with holes (or voids) situated between chains of test particles. It was in pancakes where they thought that galaxies would form. Their map showing voids and filaments was a remarkable result at the time, and it stands in marked contrast to the 1979 hierarchical simulation by Aarseth, Gott, and Turner or any other simulation done in the west. The difference can be explained in part by the Russian group's use of an intermediate-scale truncation in the adiabatic spectrum of initial irregularities, but most importantly, by the use of more realistic starting conditions defined by the Russian group. It is important to note that the Russian group had expected to see "pancakes" in 3D simulations but were surprised to see an interconnected filamentary network of matter. Upon reflection, they agreed that the result was reasonable, but no one had predicted it.

The Tartu Observatory astronomers used Shandarin's early results – having no reason to question them at the time – to guide their analysis of their sparsely sampled galaxy distributions discussed in the previous chapter. Einasto (2014, p. 126) admits that they did not know the scale of the simulation: did it represent multiple superclusters or was the web-like structure something smaller that occurred within a single galaxy cluster as the material collapsed? Jõeveer and Einasto assumed that the scale is such that the simulation includes multiple superclusters. As noted above, the Tartu Observatory group also mistakenly assumed that extended Zeldovich pancakes formed walls that separated empty voids. Because their observations of the galaxy distribution at the beginning of their studies (1977–80) were inadequate and lacked the details provided by complete redshift surveys, Jõeveer and Einasto were free to make these speculative assumptions.

The Zeldovich pancake theory was vulnerable to the same apparent problem encountered by Peebles's original hierarchical picture: point-to-point irregularities on the sky in the CMB had to be $\sim 10^{-4}$ or larger to make the original Zeldovich model work, but by the late 1970s, those who measured the CMB at radio wavelengths began to set limits on these irregularities. By the early 1980s, the limits became strict enough that neither the top-down nor the bottom-up models would work. The easiest way to fix both models was to introduce dark matter.

A new twist arose in 1980 when a group headed by Russian physicist A. Lyubimov (1980) reported that he and his group had measured the rest mass of the electron neutrino to be 30 eV (electron volts) at ITEP in Moscow.[3] There are three neutrino types: electron neutrino, mu neutrino, and tau neutrino. If the sum of all three masses exceeds ~20 eV, the mean mass density of neutrinos would be high enough to close the Universe. The exact value needed for closure depends on the Hubble constant. Although several other astrophysicists had anticipated the implications of massive neutrinos (Cowsik and McClelland 1973, Szalay and Marx 1976, Rees 1977), the Zeldovich group leapfrogged the earlier results and applied the Lyubimov result to the structure of the Universe, in general, and to the formation of galaxies, in particular, by adopting neutrino dark matter. An initial paper by Zeldovich and Sunyaev (1980) reviewed the basic characteristics of a neutrino-dominated Universe and suggested that too much mass in neutrinos might require the introduction of a cosmological constant to insure an age for the Universe consistent with astronomical observations. They also suggested that massive neutrinos would explain the hidden-mass paradox of both galaxies and clusters of galaxies.[4] Two other papers with Doroshkevich (1980a, 1980b) showed that massive neutrinos would not affect the early epoch of nucleosynthesis nor would they introduce significant irregularities in the cosmic background radiation, while they would explain the formation of large supercluster-scale structures formed by a later collapse of baryons into the neutrino-based structures. Because neutrinos move at nearly the speed of light, they quickly smooth away smaller galaxy-sized irregularities but leave intact the supercluster scale perturbations. In this sense, the models with neutrino dark matter resemble the earlier Zeldovich adiabatic models dominated by baryons with an intermediate-mass truncation in the perturbations.

To be transparent in telling this story, I must reveal that Lyubimov's result for the mass of the electron neutrino was simply wrong. Measuring the neutrino mass turns out to be a hard problem, and today – nearly four decades after the ITEP result was published – the best we have is a widely accepted upper limit for the total mass of all three neutrinos that falls in a range slightly less than 1 eV (assuming the standard LCDM model). Being unaware of Lyubimov's error, the Zeldovich group pursued this avenue of research with great speed and enthusiasm. They already had tools for modeling the collapsing gaseous pancakes, so to extend their models to incorporate hot neutrino dark matter was not a big step for them. It was in the early phase of this era that Zeldovich traveled to the IAU General Assembly meeting in the summer of 1982 (described in Chapter 5). By December 1982, Zeldovich, Einasto, and Shandarin (1982) published a review article for the prestigious British journal *Nature*, an article they entitled "Giant Voids in the Universe," that brought together all the components that they believed at that time they needed to explain the origin of galaxies and the emerging concept of the filamentary cosmic web. But once again, these theoretical ideas were simply wrong as was the claim in the 1982 paper in *Nature* that galaxies are organized in a cell structure with a "Swiss cheese" topology. The article was less a review than a summary of the flawed top-down model.

Lyubimov's 30 eV neutrino mass also gained the attention of cosmologists in the West, for example Bond, Efstathiou, and Silk (1980). In this era, the widely respected British physicist Dennis Sciama (1926–99) – PhD advisor to the current Astronomer Royal Sir Martin Rees and PhD advisor to the late Stephen Hawking – was sharing his time between Oxford, England, and Austin, Texas. At the University of Texas in 1981 Sciama finished his work with his newest PhD student named Adrian Melott (b. 1947). Upon hearing of the Lyubimov result while visiting Oxford, Melott immediately turned his attention to neutrino dark matter, with the enthusiastic support of Sciama. Melott had at first worked on possible radiative decay of the neutrinos and its effect on the gas and dust within the Milky Way, but he quickly switched to study gravitational clustering. Melott's doctoral dissertation was entitled "Massive Neutrinos as Galactic Halo Material: Radiative Decay Constraints and Gravitational Clustering." It took little to no time for Melott to begin building computer models that resembled those of Doroshkevich and Shandarin.[5] By mid-October 1981, Melott had submitted a paper to the *Astrophysical Journal* describing his effort to track 10,000 test particles in one dimension with a program based on the CIC method. The choice of one dimension was needed to resolve the internal phase-space structure of 1D "pancake" collapse to determine whether neutrinos could still be captured in galactic halos. Because of required revisions, his manuscript would not appear in the journal until 1983. This did not deter Melott from publishing two papers on his own in 1982 using the CIC method to simulate in 2D the formation of gravitational superclustering under the influence of neutrino dark matter. Melott also interested others at the University of Texas and began collaborations with Joan Centrella, Paul Shapiro, and Curtis Struck-Marcel.

Meanwhile, back in Moscow, Shandarin had begun to work with Klypin, who had joined the Doroshkevich team in 1979. By April 1982, Klypin and Shandarin completed a 3D model to test the growth of the large-scale structure and the formation of galaxies and superclusters incorporating neutrino dark matter. Before their work appeared in the journal, however, Melott applied for and received financial support for an IREX Fellowship[6] to visit Moscow State University. In the spring and early summer of 1983, Melott worked with Shandarin and Klypin in Moscow, as well as with the Tartu Observatory astronomers, to extend the 3D analysis to CDM!

Before moving to a discussion of CDM, it is best to close the massive neutrino episode. Melott, and his collaborator Centrella, submitted their own 3D simulation of galaxy formation with neutrino dark matter, and these two closely related papers – Centrella and Melott (1983) and Klypin and Shandarin (1983) – were both published at about the same time. That same year, a new group, as discussed more fully further, consisting of Frenk, White, and Davis, released their first results for a neutrino dark matter model and concluded their paper with uncertainty as to whether neutrino DM would work. These were among the last papers on dark matter consisting entirely of neutrinos. All groups had uncovered shortcomings. The 3D models lacked fine filamentary structure and displayed a galaxy distribution that was too smooth to match the structures visible in the 3D galaxy redshift maps available in 1983. Simply put, neutrino dark matter did not replicate the general character of the large-scale structure, so it became necessary to abandon neutrino-based Zeldovich pancake models even before the dubious Lyubimov 30 eV neutrino mass measurement was shown to be incorrect.

In terms of the historical development of these ideas, it is significant that the 1983 attempts by Shandarin and Klypin, by Centrella and Melott, and by Melott himself to explain cosmic voids and the large-scale structure all referenced the original redshift surveys. As discussed in the previous two chapters, this included Gregory and Thompson (1978), Einasto, Jõeveer, and Saar (1978), Tarenghi, Chincarini, Rood, Thompson, and Tifft (1981), Gregory, Thompson, and Tifft (1981), Kirshner, Oemler, Shechter, and Shectman (1981), and to the extent that it was useful, the CfA1 survey by Davis, Huchra, Latham, and Tonry (1982).[7]

7.4   Cold Dark Matter and Galaxy Formation

In 1978 – three years before the first 3D simulations were developed in Moscow – Simon White and Martin Rees (b. 1942), who were working together in Cambridge, England, introduced a new concept into galaxy formation theory. They suggested that dark halos consisting of some form of yet-to-be-identified cold, collisionless material might form first, and once these dark halos were in place, ordinary matter (the baryons) could fall into preexisting halos to form galaxies (White & Rees 1978). This is the foundation for today's CDM model of galaxy formation. Earlier calculations by Rees and Ostriker (1977) were important for the White and Rees theory of CDM. Perhaps because White and Rees made no suggestion as to what form the collisionless material might take, the concept sat dormant for several years. New interest in White and Rees's idea was ignited in 1982 when suggestions were made as to what the collisionless CDM candidate might actually be. The first of these publications was written by Bond, Szalay, and Turner (1982), followed by a second written by Blumenthal, Pagels, and Primack (1982). Finally, near the end of this same year, Peebles (1982b) worked out the basic theoretical concepts showing that CDM irregularities could explain galaxy formation in a hierarchical sense, and he referenced the two papers on CDM published earlier that year.

Adrian L. Melott (b. 1947)

Adrian Melott was minister of the Unitarian Universalist Church of Tampa, Florida, through the 1970s, when he was lured back into physics by a parishioner – and received an MA in 1977 through part-time coursework. He started graduate school at University of Texas where he met Dennis Sciama, who must rank as one of the best graduate supervisors of the twentieth century. Shortly after beginning research with Sciama, he learned of the Lyubimov experiment suggesting a 30 eV neutrino mass. With Sciama's enthusiastic agreement, he transferred his work to dark matter. As a graduate student, he performed the first N-body high-resolution simulations that analyzed the internal structure of a one-dimensional "pancake" collapse. He received his PhD in 1981 and moved to a postdoctoral position at University of Pittsburgh where he constructed the first 3D simulations of a CDM-dominated universe with 32,768 particles, which he transported to the Soviet Union while on an IREX Fellowship in 1983. Analyses performed in Estonia and Moscow showed that this model also produced a void-supercluster network. He continued his analysis of cosmological simulations, pioneering the use of supercomputers for this purpose as an Enrico Fermi Fellow at University of Chicago, then he joined the faculty of University of Kansas in 1986. He recruited Sergei Shandarin a few years later. During the 90s, Melott showed how the CDM model combined aspects of the "top-down" and "bottom-up" models to produce a supercluster-void network in the context of hierarchical clustering. Also, in a series of papers he showed that most of the claims of high resolution in numerical computer simulations are plagued by discreteness noise and two-body scattering. In 2003, Melott switched to "astrobiophysics" – considering the impact of astrophysical events on terrestrial life. He is a Fellow of the American Physical Society and the American Association for the Advancement of Science, and an Emeritus Professor of University of Kansas. Photo reproduced with permission: copyright Adrian Melott.

Because the subject of CDM was "in the air" by 1983 when Melott visited Shandarin and Klypin in Moscow, it was only natural that Melott take along his new 3D models to test side by side neutrino dark matter and CDM. This work led to a joint publication in 1983 showing the comparison. They demonstrated that CDM produced a better fit to the 3D cosmic void and supercluster redshift survey maps than neutrino dark matter. All models of structure formation are sensitive to the distribution of the initial fluctuations adopted when the calculations start. Melott and his collaborators properly used the Zeldovich Approximation to begin their simulations, and while the distribution of irregularities chosen by Melott et al. does not precisely match the preferred assumptions used today, it is a close approximation. Melott et al. (1983) has six authors: three built the 3D models and the other three provided an observational basis for testing these models. Einasto (2014, p. 160) tells how he and his colleagues at Tartu Observatory contributed.

At about the same time in 1982 and 1983 that Melott, Shandarin, and Klypin were working in Moscow, a group of Western scientists entered the scene and began to build their own 3D computer models. At the start, the new group consisted of three scientists: Carlos Frenk, Simon White, and Marc Davis. Just like Melott, Shandarin, and Klypin, at first, the new group tried to explain the filamentary distribution of galaxies with neutrino dark matter. They used the method of Aarseth to calculate interactions between particles (unlike Melott, Shandarin, and Klypin who were using the Hockney CIC method). This provided additional fine detail so the model could be compared to observations of the galaxy distribution (i.e., the galaxy correlation function). In the first of two early publications, Frenk, White, and Davis compared the old-style random initial conditions with the initial conditions from the 1970 Zeldovich Approximation first introduced into the models by Doroshkevich and Shandarin. They found the Zeldovich Approximation to be the favored choice for the model initial conditions. But in the course of their work (see Frenk, White, and Davis (1983) as well as White, Frenk, and Davis (1983)), they demonstrated that neutrino dark matter would not work to explain the galaxy distribution.

Before describing more about the new work of Davis, White, and Frenk, I briefly introduce two additional theoretical concepts that were destined to be incorporated into the new models. The first of these involves the initial spectrum (or the distribution according to size) of irregularities that seeded galaxy and supercluster formation by affecting the CDM. As described near the beginning of this chapter, in the original face-off between the top-down and the bottom-up theories, each arbitrarily selected their own form for the initial irregularities, either isothermal or adiabatic. However, in the period 1977–83, three prominent physicists – Stephen Hawking (1942–2018), Gary Gibbons (b. 1946), and Alan Guth (b. 1947) – showed how adiabatic-like irregularities can be generated from quantum fluctuations during the inflationary phase of the Universe.[8] These irregularities are now called "curvature fluctuations": irregularities in the dark matter distribution are locally compensated by opposing fluctuations in the baryons and the radiation.

Fluctuations generated by inflation are nearly scale-invariant. In other words, there are irregularities on all scales, and the irregularities are generally of the same strength independent of their linear extent. After the initial fluctuations are produced in this way, they are diminished in amplitude during the inflationary expansion. Then the larger fluctuations pass outside our horizon at early times. The shorter wavelength fluctuations reenter our horizon early on when the Universe is dominated by radiation, and they acquire larger amplitudes than fluctuations entering at later times when matter dominates over the radiation. This difference helps to explain how galaxy formation (on the smallest scales) gets an extra boost or "kick-start" in the CDM models, despite the small observed fluctuations in the Cosmic Background Radiation that were detected in the 1980s from the larger scales. I might note that in 1983 Melott et al. used a close approximation (but not the exact form) of this distribution of irregularities when they demonstrated that the CDM model provides a good fit to the observed large-scale galaxy distribution.

The second new theoretical concept is called "biasing." I discussed this idea in Chapter 2 when describing the "peak-patch" explanation for the large-scale structure in the galaxy distribution. Bias simply means that galaxies (and clusters of galaxies) will form preferentially in those regions that possess the strongest CDM irregularities. With biasing, smaller irregularities in the dark matter distribution attract fewer baryons and may never host galaxies. Kaiser (1984) introduced the theoretical concept of bias when he discussed the strong correlation between Abell rich clusters of galaxies.

George Efstathiou began to build computer models to simulate individual galaxies and clusters of galaxies at Oxford University when he was working on his PhD thesis in 1979. After graduating he applied his computer skills to several projects, and at some point, Davis, White, and Frenk invited him to join their collaboration making it a group of four (see Figure 7.2). At times, these researchers have been referred to as the "Gang of Four" because of their aggressive approach to solving the problem of the large-scale structure. When discussing their publications, I will write the four capital letters of their last names. Efstathiou brought with him a numerical method called the "particle-particle/particle-mesh code" or P3C. In 1985, and just a few months after publishing tests of the P3C computer code as applied to the CDM problem (EDWF 1985), this group of four published a widely cited paper entitled "The Evolution of Large-Scale Structure in a Universe Dominated by Cold Dark Matter" (Davis, Efstathiou, Frenk, and White 1985; DEFW). It resurrected the CDM model that was studied two

Figure 7.2 Gang of four. Photograph from 1983 at University of California
Berkeley. Left to right: Marc Davis, George Efstathiou, Simon White, and Carlos
Frenk. Copyright and photo-credit: S.D.M. White.

years earlier by Melott and his collaborators, applied the formula for curvature fluctuations from the inflationary model, and most significantly,
their computer model had a somewhat increased spatial resolution.

Significant credit needs to be given to others in the astrophysics community who also realized the potential of CDM models of galaxy formation at the time the DEFW group was writing their key papers. One of the
most prominent contributions in this regard was made by the team of
Blumenthal, Faber, Primack, and Rees (1984), who recognized the attrac-
tive and even compelling features of CDM models, especially when they
considered the full range of scales from those of individual galaxy formation through supercluster formation. These are issues not specifically
addressed in this book but are of great significance in the selection of
appropriate models to explain the origin of cosmic voids and supercluster
structure. In short, DEFW cannot be given all of the credit for the
advances that were being made in the mid-1980s to explain galaxy and
cluster formation.

At the outset, DEFW chose to work with relatively small test volumes. For
example, their 1985 CDM model with a flat "Euclidean" geometry covered
a volume corresponding to a cube 32.5 h^{-1} Mpc on a side,[9] and their initial
analysis concentrated on the nature of the galaxy distribution on scales of 1 to

10 Mpc (3 to 30 million light-years). With the improvements they had added in 1985, they could test things like the evolution of the galaxy two-point correlation function on small scales, thus moving beyond what Melott, Shandarin, and Klypin achieved in this era. A very significant result reported by DEFW is that the introduction of bias significantly improves the way the CDM model appears to match the observed characteristics of the galaxy distribution. Biasing was needed because it provides a way to reduce the relative random velocities between galaxies, as seen in their models.

Judged from afar in 1985, the new DEFW model appeared to be an unlikely concoction of disparate ideas. It relied on CDM, even though no one knew the dark matter composition. The DEFW model utilized concepts from the inflationary Universe – namely a flat Universe and the scale-free spectrum of irregularities – even when the inflationary concepts were still in a formative stage. Finally, they invoked biased galaxy formation to make their model work. At one point, Peebles (2003) called this CDM model a "house of cards."[10] For DEFW, these were significant gambles, and yet this model works even today, more than thirty years after the original DEFW publications. At least so far, no one has produced a viable alternative.

While the Gang of Four's 1985 paper, DEFW, established a necessary foundation for fitting their N-body models to the galaxy distribution, their next paper, WFDE (1987), was the one that Gregory and I found especially compelling. WFDE addresses the formation of rich galaxy clusters, superclusters, and cosmic voids under the influence of CDM in much the same way that Melott et al. did in 1983, but with several new specific results. To investigate the structures identified in their test volume in a statistically significant fashion, WFDE required new and larger models than those used in their 1985 paper. This was done in two steps. The first of these larger models placed 32,768 particles within a cube 280 Mpc (913 million light-years) on a side, and the second simulation followed 216,000 particles within a cube 360 Mpc (1.2 billion light-years) on a side. Each particle represented several galaxies. The first of these models was run on a computer twenty-five separate times (each with its own randomly generated starting conditions) to insure they could see the full range of structures that would form. The largest of their simulations pushed the limits of the computer technology of the day and was run only once. By 1987, when WFDE was published, they had abandoned the lower density models of the Universe that were tested in DEFW so that they could concentrate on the more favored flat Euclidean model, the natural choice for an inflationary Universe. With appropriate selection of the biasing parameters, this model fit the smaller-scale characteristics studied in DEFW as well as the large-scale structure results presented in WFDE. In addition, these new models were tested against the observational results like those of Koo, Kron, and Szalay (1987), who had conducted a very narrow pencil beam redshift survey that revealed extreme structures that were detected over very great distances.

From the Gregory and Thompson perspective, the WFDE paper was a welcome relief, as it confirmed theoretically what we originally found in the observations of the galaxy distribution, and was consistent with the Melott et al. (1983) analysis. With the results from the largest simulation, WFDE assembled a mock catalog of rich Abell clusters that had 27 entries, whereas in a similar true volume of the natural Universe, they would have predicted 35 rich clusters. This helped to validate the realistic nature of their models. The WFDE simulations show the "frothy" 3D filamentary structure that appears in the true galaxy distribution, and the cosmic voids in the simulations matched in size the Boötes void of Kirshner et al. at the extreme and showed others resembling those that the Arizona redshift surveys had identified. The WFDE simulations contained filament-like superclusters resembling the extended Perseus-Pisces supercluster. Rather than trying to match any statistical studies of supercluster lengths and cosmic void sizes (available only in somewhat rudimentary form in 1987), WDFE chose to match their model to the largest structures identified at that time: the Boötes void and the Perseus-Pisces supercluster.

WFDE harshly took to task those observers who tried to infer the physics of galaxy and supercluster formation based simply on the appearance of the large-scale structure. They cited three specific examples. One of these was the Tartu Observatory group, who had discussed "polyhedron cell structures" suggesting an association with the (by then defunct and simpler) Zeldovich pancake model. A second example was the CfA2 group (de Lapparent, Geller, and Huchra 1986), who had argued in favor of an explosive galaxy formation theory. The third was their own Marc Davis, who had simply claimed, when he published the CfA1 redshift survey results, that the observed galaxy distribution "presented a severe challenge to all theories of galaxy and cluster formation." While the first two criticisms were justified, the third seemed inappropriate. After all, Davis' comment in his 1982 paper was nearly an exact quote from the Gregory and Thompson 1978 cosmic void discovery paper. In both 1978 for us and in 1982 for Davis, the highly structured large-scale galaxy distribution was a challenge for the theories of galaxy and cluster formation in those earlier times. Not until the publication of the analyses of CDM models by both Melott et al. in 1983 and WFDE in 1987 did the structure in the galaxy distribution – both on the scale of galaxies and galaxy superclusters – fall under the wing of conventional models of structure formation. In fact, the dramatic features we first saw – the cosmic voids and the supercluster structure – were a great challenge until inflationary ideas inspired a redefinition of the spectrum of irregularities that allowed nearly simultaneous growth of both the small (i.e., galaxy-sized) and large (i.e., super-cluster-sized) structures. As such, both the work of Melott et al. in 1983 and the Gang of Four's work in 1987 brought to a close the major theoretical challenge posed by the discovery of the cosmic void and supercluster structure.

Those who wish to review the flow of both the theoretical and the observational advances made in this period can refer to the comprehensive timeline in Table 8.2. A bold highlight on the label "CDM" has been used to show the key theoretical papers that made advances in cold dark matter models. Similar bold highlights are shown for the contributions of the two Center for Astrophysics redshift surveys, CfA1 and CfA2. From the timeline development, the influence of the redshift survey observations on the CDM models can be quickly inferred. Notice that the CfA2 work under the title of "Slice of the Universe" came at the very close of the development period during which the CDM models were emerging. It was the galaxy structures discovered in the Arizona redshift survey work that carried the most significance in this early period. After the publication of the CfA2 Slice of the Universe, the newer survey data became the dominant influence.

I make no attempt to review subsequent developments. Not only did galaxy redshift surveys grow in size, far beyond the scope of the Arizona galaxy redshift surveys and the scope of the CfA surveys, but CDM computer simulations also began to proliferate. 1987 certainly was a time of reckoning for the naysayers and holdouts who had refused to accept the early discovery of cosmic voids and super-cluster structure. Ten years had passed since Gregory and I saw the first voids and detected the first bridge connecting two rich Abell clusters. By 1987, there appeared to be unanimous acknowledgement that a new pillar had been added to observational cosmology. The chronology of significant papers from this development can be seen in column three of the timeline in Chapter 8 (see Table 8.2).

The final topic in this chapter concerns the manner in which the top-down model of galaxy formation merged with the bottom-up model. This merger first became feasible when the introduction of the CDM model forced a redefinition of the primordial spectrum of irregularities. Instead of having "isothermal" irregularities on smaller scales and "adiabatic" irregularities on larger scales, the CDM model prescribes "isocurvature" irregularities extending over all scales. In the early models, the irregularities also seemed to require a cutoff or truncation in the adiabatic perturbation spectrum (on the low end of the mass distribution) to create the supercluster filaments. For example, Zeldovich models from the 1970s as applied to baryons used Silk damping to impose a low-mass cutoff on the perturbation spectrum. When neutrinos were suspected to be the dark matter, the cutoff was said to be caused by the free-streaming length of the neutrinos. However, as was carefully explained by Melott (1993), large-scale structure formation models that start with the ZA do not actually need

(a)   (b)

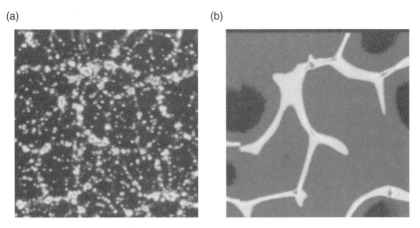

Figure 7.3 Computer simulation of galaxy and structure formation. Beacom et al. 1991 studied CDM models, all of which began with initial conditions defined by the Zeldovich approximation. By systematically varying the spectral slope of the perturbation distribution and selecting various length-scales for its truncation, they show that the resulting supercluster structure takes on a wide range of properties. The left panel (a) shows the results of a full-spectrum case (i.e., no truncation) appropriate for hierarchical galaxy formation. The right panel (b) shows an alternate simulation with an artificial low-mass cutoff imposed on the initial perturbation distribution. Notice the drastic change in the galaxy distributions between panels a and b. The "frothy" appearance of small clumps of galaxies on the left has disappeared and is replaced on the right with a high-contrast filamentary galaxy distribution. The long-wave perturbations in the initial conditions of both simulations have the same amplitude and phase. Note the similarity in the location of the major structures between the two figures. Therefore, the same physics that determines the large-scale features in the top-down model works in hierarchical clustering. The first top-down simulations, circa 1980, simply guaranteed (or even forced) the formation of filamentary structure by imposing a low-mass cutoff in the perturbation spectrum. With permission of the A.A.S.: Beacom, J., Dominik, K., Melott, A., Perkins, S. and Shandarin, S. (1991). *Astrophys. J.*, 372, pp. 351–363.

a spectrum cutoff as part of the initial conditions: large-scale structure can emerge naturally without a cutoff. As applied to models of CDM – to quote directly from the title of Melott's 1993 paper – "Peebles and Zeldovich were both right." The face-off between these two models was resolved in the most natural way possible. I show as an example the results of a computer simulation by Beacom et al. (1991) in Figure 7.3 to demonstrate that superclustering and cosmic voids can emerge from models that have no cutoff in the initial spectrum of irregularities.

8

Priority Disputes and the Timeline of Publications

With our current knowledge and understanding of dark matter and with sophisticated galaxy formation models in hand, the discovery of cosmic voids and the large-scale distribution of galaxies is simpler to grasp today than when it was unfolding in real time. Chapter 6 addressed the complexities that arose in the late summer of 1977 when Gregory and I had already submitted our first redshift survey results to the *Astrophysical Journal*, results that showed in clear detail two well-defined cosmic voids and a bridge of galaxies connecting two rich cluster cores. All the while, the Tartu Observatory group was still in the process of negotiating with the editor and referee at the *Monthly Notices of the Royal Astronomical Society* regarding their first significant publication on this topic in a conventional refereed journal. According to Einasto (2014, pp. 139–40), one of the points of discussion with the referee was whether the Russian theoretical model of structure formation would be bundled with Tartu Observatory group's somewhat limited observations. With hindsight, the astute referee was correct to object. As Chapter 7 shows, at that point in time, the early Zeldovich models were neither complete nor correct. They included no dark matter, their distribution of initial irregularities was falsely truncated at small scales, and Einasto had guessed incorrectly that the topology of the large-scale structure in the galaxy distribution was that of repeating 3D-closed cells like a honeycomb: empty voids surrounded on all sides with Zeldovich pancakes. The computer model from Shandarin that they were relying upon (see Figure 6.3) was a 2D simulation, and only much later did it become clear in 3D models that galaxy filaments were dominant, not pancake-like sheets and that the topology was like that of a sponge. The Tartu group was making speculative guesses with inadequate observational support. Their primary advantage was Shandarin's model that incorporated the Zeldovich

Approximation in the initial conditions. Gregory and I maintained a low profile in our interaction with the Tartu Observatory astronomers at that time because we did not attend the scientific meeting held in Tallinn, Estonia, in mid-September 1977. Our empirical results were impressive enough and stood on their own.

First on the agenda for this chapter is to discuss the issues posed by the redshift survey work done at the Center for Astrophysics (CfA) at Harvard University and to assess its impact on the discovery process. The Harvard work came in two stages. The CfA1 survey was initiated and led by Marc Davis (Davis et al. 1982) while the CfA2 survey was a joint effort shared between Margaret Geller and John Huchra (cf. Geller and Huchra 1989). As time has passed, both the CfA1 and CfA2 researchers have tended to call their own work "the first redshift survey" and exclude any discussion of the Arizona redshift survey publications. One key difference between our work and that from the CfA was the stated aim of each endeavor. Gregory and I formulated our observing plan from the start with the clear intent of uncovering structural features in the galaxy distribution (see Appendix A). In 1978, Davis' observations were just starting, and as reported in Section 5.6, he told me face-to-face that he was not looking for structure as the CfA1 redshift survey began. Even more extreme were Geller and Huchra, who denied the reality of cosmic voids and set out to prove that voids do not exist (Geller 1991) but ended up confirming our results. At the point where they first saw in their data the same structure we had reported and discussed, they changed course and suggested that their discoveries were original. I will also discuss how citations to scientific publications were used to enhance some parts of the discovery story and to obscure others.

The last set of issues addressed in this chapter concern the priority claims of Guido Chincarini. Chincarini and Rood had sufficient redshift survey data by 1975 to redefine the nature of the large-scale galaxy distribution in and around the Coma cluster of galaxies, but they did not use it to their advantage. In the last of their early publications before Gregory and I stepped into the fray, Chincarini and Rood (1976) suggested a model for the galaxy distribution that was still a vestige of the Hubble and Zwicky paradigm. Only after Chincarini heard a full explanation of the Gregory and Thompson cosmic void concept in the late summer 1977 in Tallinn (from Tifft's presentation) did he catch on. Then in an unprecedented move already described in Section 5.5, Chincarini took a not-yet-published and somewhat unfinished redshift sample from the Hercules supercluster region that belonged to a five-partner collaboration, and without permission from his collaborators, published a single-author paper in *Nature* discussing the data in terms of the Gregory and Thompson concept of voids. Chincarini's "rogue" paper appeared in print before both the Gregory and

Thompson and the Tartu Observatory papers were released, but because discoveries are dated based on when a paper is received by a journal, Chincarini's *Nature* paper comes in second or third even under the assumption that his use of the data set and the single author submission are considered legitimate.

Because discovery is a process – as explained by Steven J. Dick (2013) – when assessing the history, it is helpful to see the steps laid out in a timeline, so at the close of this chapter in Table 8.2, I display and then list references for the major publications. The table has three columns: CMB events in the first column, observational discoveries related to cosmic voids and supercluster structure in the middle column, and theoretical contributions to our understanding of galaxy and structure formation in the third column. Information on the CMB is included to provide historical perspective. For example, the original Penzias and Wilson (1965) manuscript is universally accepted as the discovery paper for the CMB, but it is worth remembering that Penzias and Wilson presented no evidence that the radiation they had detected possesses the spectral distribution of a black body. Evidence for the 2.7 K spectral distribution began to appear within eight months of the initial discovery (e.g., Roll and Wilkinson 1966), but it took much longer than this to fully prove the black body nature of the radiation. In every sense, Penzias and Wilson were given significant leeway in terms of what constituted in the end a very profound discovery. Perhaps their discovery was easy to accept because it meshed nicely with the earlier work by Gamow, Alpher, and Herman on the hot early Universe? Perhaps its acceptance was fast due to the persuasive nature of the companion paper by Dicke, Peebles, Roll, and Wilkinson (1965) published back-to-back with the Penzias and Wilson result? The combination of new observations with an elegant theoretical explanation was compelling.

This can be contrasted with the resistance we encountered when we first proposed that cosmic voids reside in and around supercluster structure. Of course, cosmic voids came as a surprise to Western science, and we had no immediate accompanying theoretical explanation. As the timeline in Table 8.2 shows, in Russia, theoretical models of structure formation emerged in 1976 (the same year Gregory and I got a first look at our data), but many crucial details remained incorrect and unknown in the theoretical models. Our observational discoveries were pushing into uncharted territory with some fraction of Western cosmologists uncomfortable with our findings. We were granted no leeway even remotely resembling that given to Penzias and Wilson.

Another point I address in this chapter is the relevance of our early studies of filaments in the galaxy distribution, like the bridge of galaxies connecting the Coma cluster to A1367 as well as the extended Perseus supercluster, to the grander picture that would eventually include the "Great Wall" of galaxies and the "Sloan Great Wall." Both our smaller structural features and these two gigantic structures in the galaxy distribution were eventually recognized as component parts of the more extended pattern of the cosmic web with its sponge-like topology. This issue is of significance to those who might want to understand the role of Gregory and Thompson (1978) in defining the cosmic web. The answer lies in the fact that structural features of the cosmic web are all part of a hierarchy. Gregory and I had detected smaller components of the structure, and necessarily so because our early surveys spanned smaller volumes. In this regard, it is difficult to define a specific point in time for the discovery of the extended cosmic web.

With these ideas as an introduction to several key controversies, I now describe the state of affairs as they were in the mid-1980s. By this time the Arizona galaxy redshift surveys had reported major progress, and the shallow (and therefore sparsely sampled) CfA1 redshift survey provided broad-sky coverage that was consistent with our results. Because the CfA1 survey did not reveal on its own decisive evidence for cosmic voids, the authors of the CfA1 primary publication in 1982 made clear reference to the Gregory and Thompson (1978) void discovery paper and other published Arizona redshift survey results when the first CfA1 paper by Davis and his team appeared. But as it turned out, the CfA1 survey was of significant value on its own when tests were eventually made of the early cold dark matter models by Davis, Efstathiou, Frenk, and White (1985) as well as the early tests to determine the topology of the cosmic web. The latter is discussed near the end of Chapter 9.

Even a cursory inspection of the Table 8.2 timeline reveals the relative significance or scientific role of the initial CfA2 survey results in the extended sequence of pioneering studies. The marquee paper of the CfA2 survey – de Lapparent, Geller, and Huchra (1986) – was published at the close of the pioneering period, even after theoretical studies had confirmed that cold dark matter models were the preferred choice. One can judge from Table 8.2 that the greatest relevance of the CfA2 survey was to draw full public attention to a field of investigation that had matured over a number of years. But many individuals – including scientists – who had never paid much attention to the early pioneering work began to suggest that the major discoveries somehow began with the CfA2 survey.

8.1   Redshift Survey Progress through 1985

As described in detail in Chapter 5, Gregory and I wasted no time after publishing the Coma/A1367 supercluster study to push forward with our Arizona collaborators – Tifft, Tarenghi, Chincarini, Rood – in various combinations to complete other redshift surveys: the Hercules supercluster in 1979/1980, the Perseus supercluster in 1981, the Supercluster Bridge in 1981, and then the

A2197/A2199 supercluster in 1984. In addition to our 24° sweep across Coma/ A1367, we studied the galaxy redshift distribution along two long angular surveys. The first included a 42° sweep in the southern galactic hemisphere through the Perseus supercluster (Figures 5.6a and b) and the second a 46° sweep in the northern galactic hemisphere that we called the "Hercules & A2197+A2199 Broad View." The Broad View plot appeared in the Gregory and Thompson 1984 paper on A2197 and A2199. It is reproduced here in Figure 8.1.

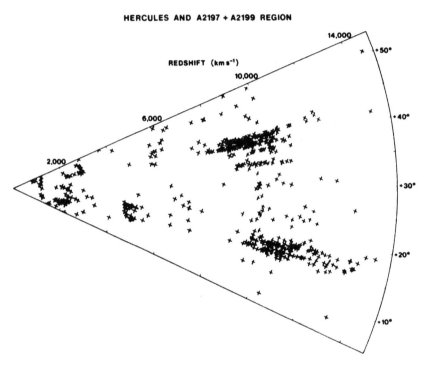

Figure 8.1 Hercules + A2197/A2199 Broad View Cone Diagram. We sit at the apex on the left and look into the Universe stretching off to the right. Each X represents a galaxy with a measured redshift. The dense group of Xs in the lower right part of the diagram (at a declination of +17° and a redshift of ~12,000 km s^{-1}) is the Hercules supercluster and the dense group of Xs sitting above it in the diagram (at a declination of +39° and a redshift of ~9,100 km s^{-1}) is the A2197/A2199 supercluster. The Finger of God redshift space distortion caused by the dynamical motion of galaxies in clusters transforms what otherwise would be dense clumps of points at the cluster location into lines pointing towards the zero-redshift origin. With permission of the A.A.S.: S. Gregory and L. Thompson 1984. *Astrophysical J.*, 286, pp. 422–36.

During this period, R. Giovanelli, M. Haynes, and G. Chincarini (1986) led the effort to use the upgraded Arecibo Telescope in Puerto Rico to complete an extensive radio-wavelength redshift survey of the Perseus supercluster. The capability of observing neutral hydrogen gas in external galaxies at 21-cm radio-wavelengths was made possible by 1974 improvements to the Arecibo telescope that included adding 40,000 individually adjustable aluminum panels to the main dish and new radio receivers. This new 21-cm survey capability provided a perfect complement to our 1981 optical wavelength Perseus redshift survey. Haynes and Giovanelli (see Figure 8.2) had appointments at Cornell University and therefore worked directly with the telescope, thereby insuring high-efficiency operation and excellent scientific results. At optical wavelengths, we detected redshifts of many elliptical and S0-type galaxies in Perseus/Pisces, and in doing so, Gregory, Tifft, and I proved in 1981 that the 2D filament that had been recognized for many years (Bernheimer 1932) is truly a spatially confined filament when studied in 3D. The Arecibo telescope added new diversity to the investigation by contributing new redshifts for spiral and irregular-type galaxies. It would be a lengthy diversion to recount here the interactions between the astronomers who were involved in the 21-cm work. Chincarini teamed at the first with Giovanelli and Haynes, but this couple continued on their own with other collaborators (but not Chincarini) over a number of years. I worked as a visiting observer at Arecibo, too, along with radio astronomers Trinh Thuan (University of Virginia) and Thomas Bania (now at Boston University), but we eventually contributed our redshift measurements to the comprehensive investigation of Giovanelli and Haynes and ended any form of competition on that front.

By 1985, the picture of the nearby Universe we were assembling was impressive and consistent throughout: cosmic voids were present in every deep redshift survey we completed, and they occupied a large fraction of the volume. The longest contiguous object studied up to that point was the Perseus supercluster that we traced in 1981 over a length that exceeded 115 million light-years (35 Mpc) while Giovanelli and Haynes at Arecibo Observatory traced it several years later to nearly twice this length: to at least 160 million light-years (50 Mpc). Along the way, Gregory and I had proven that every nearby rich Abell cluster core was embedded in its own extended supercluster structure connected to its closest neighbors. In total, this represented eight to nine years of hard work starting with our first redshifts in the Coma/A1367 region that we began to collect in early 1976. In addition to having made these observational discoveries, we watched as N-body computer simulations matured through the early 1980s repeatedly referencing our work. Together, like a hand in a glove, these studies were collectively making a profoundly significant contribution to galaxy formation theories and to cosmology.

Figure 8.2 Arecibo radio telescope. Martha Haynes and Ricardo Giovanelli, astronomers associated with Cornell University, are shown standing in the immediate foreground of the giant Arecibo radio dish. The far rim of the smooth extended spherical surface is visible just below their shoulder level. The 305 m diameter dish is fixed in a bowl-shaped depression in the hilly countryside near Arecibo, Puerto Rico. The radio receiver platform, suspended on cables far above the reflective surface, can be seen in the upper left corner of the photograph. Hanging below the platform assembly is a rotating curved track, and on this track are interchangeable long spear-shaped antennas that accept the incoming radio waves. The antenna of choice (selected according to the wavelength being studied) is moved slowly along the rotating curved track to compensate for the Earth's rotation, so the radio waves from a single celestial object can remain focused on the antenna as the object slowly moves overhead across the sky due to the Earth's rotation. Reproduced with permission: copyright Martha Haynes.

8.2   East Coast Recalcitrance and the Trouble that Followed

There had been two major pockets of resistance to our new view of the galaxy distribution. The most hard-lined resistance came from Peebles at Princeton University. He held tight to the concept he had published with Soneira in 1978 bringing into question the reality of filamentary structure in the galaxy distribution by suggesting it might be a figment of the imagination, a tendency of the eye to connect unrelated points in random distributions of galaxies. The second pocket of resistance came from Harvard University's

Center for Astrophysics where Margaret Geller and the late John Huchra were working on other projects (i.e., not large-scale redshift surveys) with the telescope and spectrograph they had inherited after the completion of the first Center for Astrophysics redshift survey (CfA1). As described in Chapter 5 (section 5.8), G. Bothun worked as a young PhD researcher in their group and watched as Geller and Huchra read about our redshift survey work in the early 1980s. Geller and Huchra (but not Bothun) rejected our concepts and doubted the reality of the structures we had discovered.

In 1985, Geller and Huchra entered the redshift survey business (Huchra actually re-entered having been part of the CfA1 survey) by collecting standard optical-wavelength redshifts in the first stripe in their new CfA2 survey. With graduate student Valerie de Lapparent, Geller and Huchra (1986) confirmed the presence of cosmic voids and supercluster structure, totally consistent with our discoveries of the previous eight years. They selected for the title of their paper "A Slice of the Universe." The results from their first slice were no surprise to us because, from their total survey length of 117°, the central 24° overlapped essentially 1:1 with the Gregory and Thompson Coma/A1367 Supercluster study. We had selected a region for our 1978 redshift survey that contained the most interesting structure: cosmic voids, the core of the Coma cluster, and two filaments branching from it, with one of these being the supercluster bridge between Coma and A1367. It was in this central region that the Coma cluster "stickman" caricature was first seen by Geller and her collaborators. Our 1978 observations had already clearly defined the stickman's body and his "western arm and leg."

However, when de Lapparent, Geller, and Huchra published their results in 1986, their paper contained no reference to our 1978 cosmic voids discovery paper. They had incorporated into their study ~300 galaxy redshifts (~25% of their total 1,100 galaxy sample) that had formed the basis of our study (obtained collectively by Chincarini, Rood, Tifft, Gregory, and Thompson), but they gave no reference as to the source of this portion of their data. The paper simply states that among the remaining redshifts, 327 have been published by other groups, but they did not specify where they found them. With this language, they avoided making any direct reference to the 1978 Gregory and Thompson study of Coma/A1367. Finally, they presented their work as though all of the structure they had detected was an original discovery. They discussed an extended pattern in the galaxy distribution consisting of large bubble-like features with sharply defined walls. Their bubbles enclosed the empty regions we had already identified as cosmic voids. They speculated that their bubble-like features could be the result of an explosive theory of galaxy formation, a concept that eventually was soundly rejected. To have excluded any reference to our 1978 discovery work ran contrary to scientific standards, especially since they had used our observations in their analysis. Their new and deeper redshift survey actually had confirmed our 1978 results, but the issue of confirmation was never raised in their paper. The editor of the journal where their manuscript was submitted for publication in late 1985, as well as the referee appointed by the editor, should have recognized this fact and should have forced a discussion of our earlier work. Our 1978 paper was not obscure. It was widely discussed, and by late 1985 it had already been cited in more than 130 other research publications.[1]

Taking a cue from the public relations fanfare that accompanied the discovery of the Boötes void in 1981, Geller and Huchra publicly discussed their own work in 1986 with a far-flung public relations initiative that led to front-page coverage in the *New York Times*, coverage in *Time* magazine, and even an interview with Geller on the popular television show *Good Morning America*. As with all public information, some was relatively sane (e.g., the *New York Times* article), but other events got out of hand. Gregory happened to have heard the *Good Morning America* segment and reported that Geller said that she had, herself, discovered voids.

Generally, Geller and Huchra were rarely in a position where our work and theirs was compared one to one. When discussing the discovery of cosmic voids, they often gave credit for this discovery to the 1981 Boötes void paper by Kirshner, Oemler, Shechter, and Shectman. If they included a reference to any work by Gregory and Thompson, the references would start with our 1981 Perseus supercluster study. It shares the 1981 publication year with the Boötes void paper. I recently surveyed all scientific papers where both Geller and Huchra are coauthors (including multiple-author papers with both Geller and Huchra listed as coauthors), and in the 24-year period from 1986 through 2010 when John Huchra died, the Geller and Huchra team referenced our 1978 Coma/A1367 Supercluster paper a single time. This one paper was published in 1999, 13 years after the "Slice of the Universe" was published. It is entitled "The Updated Zwicky Catalog (UZC)" by Falco, et al. (1999), and it contains catalogued data of new as well as previously published galaxy redshift observations. It includes no scientific interpretation or discussion. In a general sense, Geller and Huchra worked together for 24 years and managed to avoid all references to our 1978 cosmic void discovery paper.[2]

In preparation for writing this chapter, I obtained a published copy of an interview of Margaret Geller that was printed in the introductory-level university textbook entitled *Realm of the Universe* (Abell, Morrison, and Wolff 1994), where Geller is featured as a role-model scientist. Prefacing the printed interview is a thumbnail sketch of her educational and family background. The interview was conducted in late 1990 or early 1991. While it was my intention to reprint a portion of this interview in order to make it easily accessible, I was unable to secure the copyright clearance to do so. Instead, I will simply paraphrase some of Geller's responses and discuss them briefly. This interview originally came to my attention when, as a university professor, I needed to select a textbook to teach undergraduate students in the mid-1990s. From the time when I first saw the interview through today, I have been very uncomfortable with the statements this interview contains. The primary author of the textbook, the late George Abell, died in 1983, and the publisher found two additional authors to edit and update his textbook in the early 1990s. Neither of these new authors was sufficiently informed about the detailed history of redshift surveys to check the veracity of Geller's statements, so she was free to say what was on her mind. Before he died Abell was a world expert in this area of research, and he was well aware of the priority in the early pioneering redshift survey work by those of us from Arizona. Abell's premature death, however, provides us with the special opportunity to learn Geller's own view of her research circa 1991 without his editorial intervention. The reference used in this book for the interview is Geller (1991). Those who wish to read the interview in full will need to find a copy of the 1994 college textbook listed in the references as Abell, Morrison, and Wolff (1994).

The most insightful question and answer is near the start of the interview when Geller was asked to name her most important scientific contribution. In her reply, she carefully shares credit with her collaborator John Huchra and states that together they discovered "big dark regions that we called 'voids.'" The stark simplicity of the quoted reply leaves no doubt that Geller was taking credit in 1991 for discovering cosmic voids. This remarkable claim is consistent with the *Good Morning America* interview Gregory happened to have heard in 1986. Geller went on to say in the 1991 textbook interview that the voids are surrounded by galaxies located on very thin surfaces, and that this phenomenon cannot be explained by any theory.

Today, the manner in which galaxies appear in sharp features at the outer edges of voids can be explained by simple linear gravitational theory (as discussed in Chapter 9, see especially Figure 9.2). The original de Lapparent et al. (1986) paper made an erroneous big deal of these sharp edges by referencing "the explosive galaxy formation theory" (Ostriker and Cowie 1981), an idea that lost its central importance rather quickly. The somewhat larger patterns in the galaxy distribution recognized by Geller, also mentioned in this interview, are discussed later in this chapter.[3]

In her interview, Geller expressed great surprise with the observations that came from the "Slice of the Universe" redshift survey (de Lapparent, Geller, and

Huchra 1986) and said that "nobody expected such a striking pattern." The level of surprise, however, depends on one's perspective. Certainly, she was surprised, because she had previously denied the significance of the redshift survey results that had been published during an eight-to nine-year period from the ongoing Arizona redshift survey work. Gregory and I were among those who were not surprised. Neither were Zeldovich and Shandarin.

I have identified only once when Geller was put in a position where she compared our Coma/A1367 Supercluster study with her Slice of the Universe. This happened in the year 2000 when author Ken Croswell was writing his book entitled *The Universe at Midnight*. He provided to me and to Gregory a forum to air our objections regarding the way Geller and Huchra were treating our discoveries. Croswell starts by noting that the late Allan Sandage had commented on Geller's lack of transparency when she had confirmed our results but had not acknowledged it. The following is reproduced from Croswell (2001, p. 138) with permission:

> Allan Sandage decried this "rewrite of the history" as a "travesty of justice," comparing Steve Gregory and Laird A. Thompson to Ralph Alpher and Robert Herman – the two scientists who predicted the cosmic microwave background, only to find others claiming credit when it was confirmed.

Then Croswell's book goes on:

> Geller countered that her 1986 work was distinctive because it swept over a wide swath of sky, included far more galaxies – 1,099 versus Gregory and Thompson's 238 – and studied regions not previously known to have unusual structures, thereby better sampling the universe at large.

I can easily show that Geller's quoted retort was a misjudgment, but before doing so, I summarize in Table 8.1 the basic facts. During the six-year period 1978–84, before de Lapparent, Geller, and Huchra began their 1986 "Slice of the Universe" study, we had extended our redshift surveys beyond the 238 galaxy redshifts we reported in our Coma/A1367 redshift survey. The total number of redshifts in our combined Arizona surveys by 1984 approached the sample size Geller and her collaborators obtained for the Slice of the Universe in 1986. As Table 8.1 shows, we had surveyed nearly three times the area (2,078 square degrees on the sky versus 702 square degrees for the Slice of the Universe), we had sampled broad areas in both the northern and southern galactic hemispheres, and the combined length of our angular sweeps was $24° + 42° + 48° = 114°$ (Geller's single slice covered a continuous sweep of $117°$). Our clearly stated goal – after our initial discoveries in the Coma/A1367 region – was to test whether all of the very richest nearby Abell clusters were embedded in an extended bridge-like supercluster structure and were thereby connected to one another. By 1984, our work had already proven this statement to be true to within the limits of our observations. Based on the facts given in Table 8.1, I assert that the sum-total of *our* redshift surveys provided better sampling of the Universe at large. The need to sample different parts of the local Universe is a significant advantage in our favor. At that time, de Lapparent, Geller, and Huchra had sampled only one strip across the northern galactic hemisphere. There was no reason for Geller and Huchra to imagine that they had done a definitive survey without looking in other regions, and furthermore, there was no basis for them to claim that they were doing original work. The primary things they did differently (and better than we did) were to display their work on the front page of the *New York Times* and in general to use the press to discuss their scientific results.

Table 8.1 *Survey areas of the pioneering Arizona redshift research*

Survey Name	Reference	Longest Angular Sweep	Sample: Number of galaxies	Survey Area: Square Degrees
Coma/A1367	Gregory and Thompson (1978)	24°	238	252
Hercules	Tarenghi, Tifft, Chincarini, Rood, and Thompson (1979)	7°	191	28
Perseus	Gregory, Thompson, and Tifft (1981)	42°	141	482
Hercules Bridge	Chincarini, Thompson, and Rood (1981)	17°	44	332
A2197/ A2199	Gregory and Thompson (1984)	12°	136	72
Hercules+A2197/ 2199 Broad View	Gregory and Thompson (1984)	48°	371	1344
TOTAL ARIZONA SURVEY AREA (through 1984)		114°	750	2078[*]
CfA2 Slice of the Universe	de Lapparent, Geller, and Huchra (1986)	117°	1099	702

[*] To be fair, in listing the Arizona total redshift survey area, I include only 3 numbers: Coma/A1367, Perseus, and the A2197/2199+Hercules Broad View from Gregory and Thompson (1984). The narrower but deeper survey areas from three other surveys (A2197/2199, Hercules, and the Hercules Bridge) are all contained within the Gregory and Thompson 1984 Broad View survey area. The same accounting is applied to the number of redshifts. The Broad View survey was not a 100 percent complete redshift survey but contained sub-regions that were essentially complete surveys.

When the late Allan Sandage drew a connection between Gregory and Thompson and the unfortunate treatment dealt to Alpher and Herman (events that occurred in 1965 when the CMB was discovered), he raised an older issue that directly involved a young P. J. E. Peebles, who had repeated and somewhat extended the work that was originally done by CMB pioneers Alpher and Herman.[4] A careful reader will recall that Peebles was Geller's PhD thesis advisor, and by 1986 when the Slice of the Universe redshift survey results were published, he had become one of the top theoretical cosmologists in the world. As it turns out, Peebles further exacerbated the cosmic void discovery issue, at least from our perspective, when, in 1993, he published his beautifully comprehensive book entitled *Principles of Physical Cosmology*. In the introduction to Peebles' Chapter 3 in a section called "Mapping the Galaxy Distribution," Peebles describes the development of and the status of galaxy redshift surveys, circa 1992. In this discussion, Peebles excluded any mention of our Coma/A1367 supercluster study and all of the pioneering redshift survey work that I detail in Table 8.1. Instead of citing the Coma/A1367 study for the discovery of cosmic voids, Peebles' book makes the incorrect statement "Kirshner et al. (1981) named these regions voids." The fact is, Gregory and I discovered cosmic voids in 1978, and we named them "voids" at that time. Unfortunately, graduate students and other researchers who would rely on his knowledge and leadership for many years were left with an incomplete version of the discovery story of cosmic voids. His book is inconsistent with the astronomy review article by Oort (1983) entitled "Superclusters." It seems that Peebles and Oort rarely saw eye to eye on this topic.

Revealing the large-scale structure in the galaxy distribution involved two steps. The first was to recognize and then to show that Hubble's and Zwicky's assumption of local homogeneity was wrong and that the galaxy distribution is dramatically inhomogeneous on scales dominated by superclusters (positive density enhancements) and cosmic voids (negative density perturbations). Gregory and I published this decisive result in 1978; we did so on a scale of 300 million light-years, that is, the depth of our Coma/A1367 redshift survey. But as we completed one survey after the next (see Table 8.1), we too recognized more and more structural irregularities as we extended our surveys and searched for bridges of galaxies that connect superclusters. This was the second step in the discovery process. So now I briefly discuss this second step in the discovery process.

As Geller and Huchra (1989) noted in the early days of their redshift survey work, the more extended the survey volume, the farther "extended patterns" could be traced in the galaxy distribution. Of course, the "Slice of the Universe" made a significant contribution because the patterns in the galaxy distribution appeared in a prominent way in their somewhat larger redshift map. However, the structure Gregory and I reported in 1978 and the structure that Geller saw on larger scales in 1986 (at the close of the pioneering period) appear to be part of a continuous distribution of irregularities. The statistical properties of this structure were discussed in a comprehensive theoretical review of the large-scale structure by John Peacock (b. 1956) of the University of Edinburgh, who showed a continuity in the structure up to the limits he could reliably measure. Peacock (2003) used, as the basis of his analysis, the 2dF Galaxy Redshift Survey. The 2dF-GRS was the first of several redshift surveys that immediately followed the Geller and Huchra CfA2 survey as astronomers pushed the observations one step at a time deeper into the Universe. The Geller and Huchra team completed the CfA2 survey by making available in ~1999 what they called the "ZCAT." This is essentially the same data set published under the title "The Updated Zwicky Catalogue" (Falco et al. 1999). It is a set of 12,925 galaxy redshift observations that is 98% complete and based on the 13,150 galaxy sample originally defined in the Zwicky "Catalogue of Galaxies and Clusters of Galaxies." This can be compared with the 2df-GRS from 2002 with a much larger sample of 221,414 good-quality spectra. While the CfA2 survey reported the discovery of the Great Wall of galaxies, the 2df-GRS clearly resolved the even larger Sloan Great Wall (that was first reported by Gott et al. 2005). As mentioned, these features are each part of a continuous distribution of irregularities.

Geller's work was one in a succession of many such studies, and the de Lapparent et al. (1986) redshift survey was neither the first nor the last in the series (see Table 8.2). Everyone is free to define their own threshold of discovery for the extended patterns in the galaxy distribution. Sergei Shandarin told me that he and Zeldovich could tell exactly what was going on in terms of large-scale structure models when they first saw the Gregory and Thompson Coma/A1367 redshift survey map in 1978. On the other hand, Peebles waited for eight more years – for the Slice of the Universe redshift map – before he could bring himself to affirm the existence of significant structure in the galaxy distribution. Following the suggestions of Dick, this must be designated as an extended discovery process. However, a non-biased judge of the cosmology community would certainly acknowledge the sharp contrast between the extended discovery process of the large-scale structure and the extremely brief discovery process of the cosmic microwave background. No comparison could be more extreme. The most conservative hold-out for homogeneity in the large-scale structure, P. J. E. Peebles of Princeton University, waited eight years to acknowledge significant structure in the large-scale galaxy distribution even though he was among the first to acknowledge that the random noise found by Penzias and Wilson in the Bell Lab's microwave antenna was evidence of the hot early Universe.

8.3   The Chincarini Challenge

Finally, I address three Chincarini papers published in 1975, 1976, and 1978. On the first two of these, Rood was a coauthor. Rood was offered coauthorship on the third paper, but he declined (Chincarini 2013). These papers were already discussed once in Chapter 5. In their clearly stated effort to find the outer edge of the Coma cluster, Chincarini and Rood traced the galaxy distribution sufficiently far from the center of the Coma cluster that, without recognizing what was happening, they began to detect galaxies that we now know to be members of the bridge of galaxies called the "Great Wall," and they probed regions where cosmic voids were present. In their analysis of the observations, however, they held tight to their method of plotting each galaxy's redshift as a function of its radial distance from the center of the Coma cluster. This method of analysis not only kept their attention diverted from a true 3D study, but their manner of displaying their data scrambled the information that is available to map the galaxy distribution. Two years later, Gregory and I completed our redshift survey, did the analysis in 3D, and produced the first true wide-angle map based on a complete redshift survey. It showed voids and supercluster structure, and then we stated for the first time the new paradigm where empty cosmic voids – distinct physical entities – sit by themselves adjacent to supercluster filaments (the filament in this case is the bridge of galaxies that Gregory and I discovered connecting Coma with A1367).

Those who suggest that the raw data from the Chincarini and Rood 1976 paper might have been sufficient to count as the discovery of cosmic voids must acknowledge (as I point out in Chapter 5) that Chincarini and Rood never gave a proper description anywhere in their 1975 or 1976 papers for the new "void paradigm."[5] They never discussed cosmic voids as astrophysical objects. Instead, they stuck with the old model of Hubble and Zwicky where clusters are positive density enhancements that sit amongst field galaxies. Then they accepted their own erroneous size for the Coma cluster claiming that it extends to a radius of nearly 12.5° (they were detecting the Coma/A1367 bridge, instead) and used their observations to calculate a low value for the density of so-called field galaxies. This led to a set of confusing (and somewhat contradictory) statements in Chincarini and Rood (1976) that provides their best interpretation of the observations. This included a statement that supercluster structures fade into the background and leave "little if any space between them." Then they mention a "pronounced effect of 'segregation in redshifts'" without offering an explanation of what it meant.

This characterization by Chincarini and Rood (1976) does not constitute the discovery of cosmic voids as separate structural entities, and it sharply contrasts with the Gregory and Thompson (1978) description where we identify specific cosmic voids, assign diameters to them and calculate the mean density in the interior volume of the voids. Because of these facts Sandage (1987) said:

> The convincing data and power visualization was by Gregory and Thompson (1978, their Fig. 2). This paper marks the discovery of voids, which have become central to the subject [of the large-scale structure]. Prior work by Einasto et al. (1980 with earlier references), Tifft and Gregory (1976), and Chincarini and Rood (1976), foreshadowed the development, but the Gregory and Thompson discovery is generally recognized as the most convincing early demonstration.

I close this chapter by saying that every research group, and every individual researcher, handles interactions with their colleagues and competitors in a variety of styles. This chapter highlights two interactions that went awry, but there were other scientists in this era who participated in the pioneering redshift survey studies, and all the while they handled the situation gracefully. Those who should be especially commended in this regard include Herb Rood, William Tifft, Massimo Tarenghi, Martha Haynes, Ricardo Giovanelli, and of course my close collaborator Stephen Gregory.

Table 8.2 *Timeline of discovery*

	CMB discovery		Large-scale structure observations		Galaxy formation & structure formation theory
	Precursors		**Precursors**		
1941	A. McKellar Temperature from CN absorption lines.	1784	Herschel: "Strata" of nebulae in 2D galaxy distribution		
		1811	Herschel: Blank regions in the 2D galaxy distribution	1937	Holmberg & Lundmark: binary galaxies grow a hierarchy
1948	Gamow, Alpher, Herman Predict 5° K CMB	1932	Shapley & Ames: local gradients & blank regions in 2D		
		1956	Humason, Mayall, & Sandage: summary of ~800 galaxy redshifts known at this time Mayall		
		1960	Electronic photography for spectroscopy		
		1965	de Vaucouleurs 3D map of Local Supercluster	1968	Peebles & Dicke: globular clusters - galaxy building blocks

(*cont.*)

Year	Event	Year	Event
		1970	Harrison: initial scale-free adiabatic irregularities
		1970	Zeldovich: 3D collapse & Zeldovich Approximation
		1970	Peebles & Yu: define CMB-based density fluctuation limits
		1972	Zeldovich: initial scale-free adiabatic irregularities
		1972	Sunyaev & Zeldovich: top-down supercluster pancake model
1975	Chincarini & Martins: Seyfert's Sextet segregation of redshifts		
1975 [ ]	Chincarini & Rood: Coma cluster halo data-no discussion	1975	Shandarin: 2D simulation with holes & filaments (to Einasto)
1976 V	Tifft & Gregory: Coma survey, foreground "devoid" of galaxies	1976	Shandarin: 2D simulation published by Doroshkevich et al.
1976 [ ]	Chincarini & Rood: Coma cluster halo & segregation of redshifts		
1957	Shmaonov CMB detection in Russia		
1964	Doroshkevich & Novikov Predict 1 degree K CMB		
1965	Dicke, Peebles, Roll, & Wilkerson Work toward CMB detection		

(cont.)

Table 8.2 (cont.)

	CMB discovery		Large-scale structure observations		Galaxy formation & structure formation theory
Detection					
1965	Penzias & Wilson: CMB detected at 7.35 cm wavelength	1977	Tallinn Meeting at Tartu Observatory	1977	Gibbons & Hawking: quantum origin of initial irregularities
			Detection		
		1978 V	Gregory & Thompson: voids discovered in Coma/A1367 3D map	1978	Doroshkevich & Shandarin: publish 2D simulation - baryons only
		1978	Chincarini's dubious paper in *Nature*: Lack of field galaxies in Hercules data	1978	Soniera & Peebles: "human eye connects random dots"
		1978 V	Joeveer, Einasto, Tago: large-scale structure consistency argument	1978	White & Rees Foundational **CDM** concept
		1979 V	Tarenghi, Tifft, Chincarini, Rood & Thompson: Hercules supercluster	1979	Aarseth, Gott & Turner: simulation with voids but no filaments

(cont.)

Year	Event	Year	Event
1966	Roll & Wilkinson CMB detected at Wavelength = 3.2 cm		
1980	Tarenghi, Chincarini, Rood, & Thompson Hercules supercluster analysis	1980	Peebles: book "Large-Scale Structure of the Universe"
1980 V	Einasto, Joeveer, & Saar Large-scale structure overview: catalog data		
1981 V	Gregory, Thompson, & Tifft Perseus supercluster redshift survey		
1981 …	Kirshner, Oemler, Schechter, & Shectman Boötes void redshift survey		
1981 V	Chincarini, Thompson, & Rood Supercluster Bridge between Rich Clusters		
1982 V	Davis, Huchra, Latham, & Tonry – **CfA1** All-sky complete: 2400 redshift sample	1982	Bond et al., Blumenthal et al., and Peebles: **CDM** theory
1982	Zeldovich, Einasto, & Shandarin *Nature* Review: neutrino dark matter	1983	Centrella & Melott: massive neutrino 3D simulation
		1983	Klypin & Shandarin: massive neutrino 3D simulation

(cont.)

Table 8.2 (*cont.*)

CMB discovery	Large-scale structure observations	Galaxy formation & structure formation theory
		1983 Melott et al.: First **CDM** 3D simulation
	1983 … Chincarini, Giovanelli, & Haynes Radio redshifts in Coma/A1367 bridge	
	1984 V Gregory & Thompson A2197+A2199 redshift survey + Broad view	1984 Kaiser Galaxy biasing theory
		1984 Blumenthal, Faber, Primack & Rees: **CDM** & galaxy formatio
	1985 V Giovanelli & Haynes Pisces-Perseus Supercluster radio redshifts	1985 DEFW: **CDM** 3D model with inflation & biasing
	1986 [] Giovanelli, Haynes, & Chincarini Pisces-Perseus Supercluster morphology	
	1986 V de Lapparent, Geller, & Huchra – **CfA2** A Slice of the Universe redshift survey	
1969 Conklin CBM shows Earth's motion	1986 Gott, Melott, & Dickinson 3D topology is sponge-like	1987 WFDE: **CDM** 3D model with superclusters & cosmic voids

(*cont.*)

1984	Uson & Wilkinson CMB early irregularity limit	1989 V	Geller & Huchra – **CfA2** Mapping the Universe: Science Magazine
1990	Mather COBE FIRAS spectrum	1993 V	Wegner, Haynes, & Giovanelli Pisces-Perseus supercluster radio redshifts
1992	Smoot COBE DMR fluctuations	1996 V	Shectman et al.: Las Companas LCRS 26,000 redshifts
2003	Spergel et al. WMAP anisotropies	1999 V	Falco, Kurtz, Geller, Huchra, Peters, et al. **CfA2** survey 18,000 redshifts
		2001 V	Colless et al.: 2dF Galaxy Redshift Survey 220,000 redshifts
		2002 V	Strauss et al.: Sloan Digital Sky Survey 700,000 redshifts
		1991–1995	Melott, Beacom et al. unify top-down & bottom-up theories
		1996	Bond, Kofman & Pogosyn: cosmic web, adhesion & biasing

Note: **Redshift** Plot Types – [] indicates a rectangular plot of redshift vs. position V indicates a cone or wedge diagram—

CMB precursor references:

1. McKellar, A. (1941). Molecular Lines from the Lowest States of the Atomic Molecules Composed of Atoms Probably Present in Interstellar Space. *Pub. Dominion Astrophys. Observatory (Victoria)*, 7, pp. 251–72.
2. Alpher, R. and Herman, R. (1948). Evolution of the Universe. *Nature*, vol. 162, pp.774–5.
3. Shmaonov, T. (1957). Ph.D. Thesis. Also reported in Pribori i Tekhniika Experimenta (in Russian), 1, p. 83.
4. Doroshkevich, A. and Novikov, I. (1964). Mean Density of Radiation in the Metagalaxy and Certain Problems in Relativistic Cosmology. *Soviet Physics Doklady*, 9, pp. 111–4.
5. Dicke, R., Peebles, P., Roll, P., and Wilkinson, D. (1965). Cosmic Black-Body Radiation. *Astrophys. J.*, 142, pp. 414–9.

CMB detection references:

1. Penzias, A. and Wilson, R. (1965). A Measurement of Excess Antenna Temperature at 4080 MHz *Astrophys. J.*, 142, 419–21.
2. Roll, P. and Wilkinson, D. (1966). Cosmic Background Radiation at 3.2 cm – Support for Cosmic Black-Body Radiation. *Phys. Rev. Lett.*, 16, pp. 405–7.
3. Conklin, E. (1969). Velocity of the Earth with Respect to the Cosmic Background Radiation. *Nature*, 222, pp. 971–2.
4. Uson, J. and Wilkinson, D. (1984). Small-Scale Isotropy of the Cosmic Microwave Background at 19.5 GHz. *Astrophys. J.*, 283, pp. 471–8.
5. Mather, J. et al. (1990). A Preliminary Measurement of the Cosmic Microwave Background Spectrum by the Cosmic Background Explorer (COBE) Satellite. *Astrophys. J.*, 354, pp. L37–L40.
6. Smoot, G. et al. (1992). Structure in the COBE Differential Microwave Radiometer First-Year Maps. *Astrophys. J.*, 396, pp. L1–L5.
7. Spergel, D. et al. (2003). First-Year Wilkinson Microwave Anisotropy Probe (WMAP) Observations: Determination of Cosmological Parameters. *Astrophys. J.*, 148, pp. 175–94.

LSS precursor references:

1. Herschel, W. (1784). Account of Some Observations Tending to Investigate the Construction of the Heavens. *Phil. Trans.*, 74, Section XIII, pp. 437–51.
2. Herschel, W. (1811). Astronomical Observations Relating to the Construction of the Heavens, Arranged for the Purpose of a Critical Examination, the Result of which Appears to Throw Some New Light upon the Organization of the Celestial Bodies. *Phil. Trans.*, Section LXIII, pp. 437–51.
3. Shapley, H. and Ames, A. (1932a). *Annals of the Astronomical Observatory of Harvard College*, 88, No. 2, pp. 43–75 (Shapley-Ames Catalogue); 1932b, Harvard College Observatory, Bulletin No. 887, pp. 1–6.

4. Humason,M., Mayall, N., and Sandage, A. (1956). Redshifts and Magnitudes of Extragalactic Nebulae. *Astron. J.*, 61, pp. 97–162.

5. Mayall, N. (1960). Advantages of Electronic Photography for Extragalactic Spectroscopy. *Ann. Astrophys.*, 23, pp. 344–59.

6. de Vaucouleurs, G. (1965). Nearby Groups of Galaxies. Ch. 14, pp. 557–600 in *Galaxies and the Universe*. eds. Sandage, A., Sandage, M., Kristian, J., and Tamman, G. University of Chicago Press (Chicago, Illinois). The Compendium Series was published in 1975 but a note in the article gives 1965 as the date of submission.

7. Chincarini, G. and Martins, D. (1975). On the "Seyfert Sextet," VV 115. *Astrophys. J.*, 196, pp. 335–7.

8. Chincarini, G. and Rood, H. (1975). Size of the Coma Cluster. *Nature*, 257, pp. 294–5.

9. Tifft, W. and Gregory, S. (1976). Direct Observations of the Large-Scale Distribution of Galaxies. *Astrophys. J.*, 205, pp. 696–708.

10. Chincarini, G. and Rood, H. (1976). The Coma Supercluster – Analysis of Zwicky-Herzog Cluster 16 in Field 158. *Astrophys. J.*, 206, pp. 30–7.

LSS detection references:

1. Gregory, S. and Thompson, L. (1978). The Coma/A1367 Supercluster and Its Environs. *Astrophys. J.*, 222, pp. 784–99.

2. Chincarini, G. (1978). Clumpy Structure of the Universe and General Field. *Nature*, 272, pp. 515–6. (The "rogue" paper.)

3. Joeveer, M., Einasto, J., and Tago, E. (1978). Spatial Distribution of Galaxies and Clusters of Galaxies in the Southern Galactic Hemisphere. *Mon. Not. Royal Astron. Soc.*,185, pp. 357–70.

4. Tarenghi, M., Tifft, W., Chincarini, G., Rood, H., and Thompson, L. (1979). The Hercules Supercluster. *I. Basic Data. Astrophys. J.*, 234, pp. 793–801.

5. Tarenghi, M., Chincarini, G., Rood, H., and Thompson, L. (1980). The Hercules Supercluster. *II. Analysis. Astrophys. J.*, 235, pp. 724–42.

6. Einasto, J., Joeveer, M., and Saar, E. (1980). Structure of Superclusters and Supercluster Formation. *Mon. Not. Royal Astron. Soc.*, 193, pp. 353–75.

7. Gregory, S., Thompson, L., and Tifft, W. (1981). The Perseus Supercluster. *Astrophys. J.*, 243, pp. 411–26.

8. Kirshner, R., Oemler, A., Jr., Schechter, P., and Shectman, S. (1981). A Million Cubic Megaparsec Void in Boötes. *Astrophys. J. Lett.*, 248, L57–L60.

9. Chincarini, G., Thompson, L., and Rood, H. (1981). Supercluster Bridge between Groups of Galaxy Clusters. *Astrophys. J. Lett.*, 249, L47–L50.

10. Davis, M., Huchra, J., Latham, D., and Tonry, J. (1982). A Survey of Galaxy Redshifts. II. The Large Scale Spatial Distribution. *Astrophys. J.*, 253, pp. 423–45. (CfA1)

11. Zeldovich, Y., Einasto, J., and Shandarin, S. (1982). Giant Voids in the Universe. *Nature*, 300, pp. 407–13.

12. Chincarini, G., Giovanelli, R., and Haynes, M. (1983). 21 Centimeter Observations of Supercluster Galaxies – The Bridge between Coma and A1367. *Astrophys. J.*, 269, pp. 13–28.

13. Gregory, S. and Thompson, L. (1984). The A2197 and A2197 Galaxy Clusters. *Astrophys. J.*, 286, pp. 422–36.

14. Giovanelli, R. and Haynes, M. (1985). A 21 cm Survey of the Pisces-Perseus Supercluster. *I – The Declination Zone +27.5 to +33.5 degrees. Astron. J.*, 90, pp. 2445–73.

15. Giovanelli, R., Haynes, M., and Chincarini, G. (1986). Morphological Segregation in the Pisces-Perseus Supercluster. *Astrophys. J.*, 300, pp. 77–92.

16. de Lapparent, V., Geller, M., and Huchra, J. (1986). A Slice of the Universe. *Astrophys. J. Lett.*, 302, pp. L1-L5. (CfA2)

17. Gott, J.,III, Melott, A., and Dickinson, M. (1986). The Sponge-like Topology of Large-Scale Structure in the Universe. *Astrophys. J.*, 306, pp. 341–57.

18. Geller, M. & Huchra, J. (1989). Mapping the Universe. *Science*, 246, pp. 897–903.

19. Wegner, G., Haynes, M., and Giovanelli, R. (1993). A Survey of the Pisces-Perseus Supercluster. *V – The Declination Strip +33.5 deg to +39.5 deg and the Main Supercluster Ridge. Astron. J.*, 105, pp. 1251–70.

20. Shectman, S., Landy, S., Oemler, A., Jr., Tucker, D., Lin, H., Kirshner, R., and Schechter, P. (1996). The Las Campanas Redshift Survey. *Astrophys. J.*, 470, pp. 172–88.

21. Falco, E., Kurtz, M., Geller, M., Huchra, J., Peters, J., Berlind, P., Mink, D., Tokarz, S., and Elwell, B. (1999). The Updated Zwicky Catalog (UZC). *Pub. Astron. Soc. Pacific*, 111, pp. 438–52.

22. Colless, M., Dalton, G., Maddox, S., Sutherland, W., and 25 coauthors (2001). The 2dF Galaxy Redshift Survey: Spectra and Redshifts. *Mon. Not. Royal Astron. Soc.*, 238, pp. 1039–63.

23. Strauss, M., Weinberg, D., Lupton, R., Narayanan, V., and 32 coauthors (2002). Spectroscopic Target Selection in the Sloan Digital Sky Survey: The Main Galaxy Sample. *Astron. J.*, 124, pp. 1810–24.

LSS theory references:

1. Holmberg, E. (1937). A Study of Double and Multiple Galaxies together with Inquiries into Some General Metagalactic Problems with an Appendix Containing a Catalogue of 827 Double and Multiple Galaxies. *Medd. Lund Obs.*, No. 6, pp. 3–173.

2. Peebles, P. and Dicke, R. (1968). Origin of the Globular Clusters. *Astrophys. J.*, 154, pp. 891–908.

3. Harrison, E. (1970). Fluctuations at the Threshold of Classical Cosmology. *Phys. Rev. D*, 1, pp. 2726–30.

4. Zeldovich, Y. (1970). Gravitational Instability: An Approximate Theory for Large Density Perturbations. *Astron. and Astrophys.*, 5, pp. 84–9.

5. Peebles, P. and Yu, J. (1970). Primeval Adiabatic Perturbation in an Expanding Universe. *Astrophys. J.*, 162, pp. 815–36.

6. Zeldovich, Y. (1972). A Hypothesis, Unifying the Structure and the Entropy of the Universe. *Mon. Not. Royal Astron. Soc.*, 160, pp. 1P–3P.

7. Sunyaev, R. and Zeldovich, Y. (1972). Formation of Clusters of Galaxies; Protocluster Fragmentation and Intergalactic Gas Heating. *Astron. and Astrophys.*, 20, pp. 189–200.

8. Shandarin, S. 1975 (private communication: hand-to-hand transfer of the computer plot).

9. Doroshkevich, A., Zeldovich, Y., and Sunyaev, R. (1976). "Adiabatic Theory of Formation of Galaxies. In Origin and Evolution of Galaxies and Stars." OEGS Conference, pp. 65–104. (in Russian).

10. Gibbons, G. and Hawking, S. (1977) Cosmological Event Horizons, Thermodynamics, and Particle Creation. *Phys. Rev. D*, 15, pp. 2738–51.

11. Doroshkevich, A. and Shandarin, S. (1978). A Statistical Approach to the Theory of Galaxy Formation. *Soviet Astron.*, 22, pp. 653–60.

12. Soneira, R. and Peebles, P. (1978). A Computer Model Universe – Simulation of the Nature of the Galaxy Distribution in the Lick Catalog. *Astron. J.*, 83, pp. 845–60.

13. White, S. and Rees, M. (1978). Core Condensation in Heavy Halos – A Two-Stage Theory for Galaxy Formation and Clustering. *Mon. Not. Royal Astron. Soc.*, 183, pp. 341–58.

14. Aarseth, S., Gott, J., III and Turner, E. (1979). N-Body Simulation of Galaxy Clustering. *I. Initial Conditions and Galaxy Collapse Times. Astrophys. J.*, 228, pp. 664–83.

15. Peebles, P. (1980). *Large Scale Structure of the Universe* (Princeton, NJ: Princeton University Press).

16. Bond, J., Szalay, A., and Turner, M. (1982). Formation of Galaxies in a Gravitino-Dominated Universe. *Phys. Rev. Lett.*, 48, pp. 1636–9.

17. G. Blumenthal, H. Pagels, and J. Primack (1982). Galaxy Formation by Dissipationless Particles Heavier than Neutrinos, *Nature*, 299, pp. 37–8.

18. Peebles, P. (1982). Large-Scale Background Temperature and Mass Fluctuations Due to Scale-Invariant Primeval Perturbations. *Astrophys. J.*, 263, pp. L1–L5.

19. Centrella, J. and Melott, A. 1983 Three-Dimensional Simulation of Large-Scale Structure in the Universe. *Nature*, vol. 305, pp. 196–8.

20. Klypin, A. and Shandarin, S. (1983). Three-Dimensional Numerical Model of the Formation of Large-Scale Structure in the Universe. *Mon. Not. Royal Astron. Soc.*, 204, pp. 891–907.

21. Melott, A., Einasto, J., Saar, E., Suisalu, I., Klypin, A., and Shandarin, S. (1983). Cluster Analysis of the Non-Linear Evolution of Large-Scale Structure in an Axion/Gravitino/ Photino Dominated Universe. *Phys. Rev. Lett.*, 51, pp. 935–8.

22. N. Kaiser (1984). On the Spatial Correlations of Abell Clusters. *Astrophys. J.*, 284, pp. L9–L12.

23. Blumenthal, G., Faber, S., Primack, J., and Rees, M. (1984). Formation of Galaxies and Large-Scale Structure with Cold Dark Matter. *Nature*, 311, pp. 517–25.

24. Davis, M., Efstathiou, G., Frenk, C., and White, S. (1985). The Evolution of Large-Scale Structure in a Universe Dominated by Cold Dark Matter. *Astrophys. J.*, 292, pp. 371–94.

25. White, S., Frenk, C., Davis, M., and Efstathiou, G. (1987). Clusters, Filaments, and Voids in a Universe Dominated by Cold Dark Matter. *Astrophys. J.*, 313, pp. 505–16.

26. Beacom, J., Dominik, K., Melott, A., Perkins, S., and Shandarin, S. (1991). Gravitational Clustering in the Expanding Universe: Controlled High-Resolution Studies in Two Dimensions. *Astrophys. J.*, 372, pp. 351–363.

27. Bond, J., Kofman, L., and Pogosyan, D. (1996). How Filaments are Woven into the Cosmic Web. *Nature*, 380, pp. 603–6.

9

Impact of Cosmic Voids: Cosmology, Gravity at the Weak Limit, and Galaxy Formation

Gravity relentlessly sweeps matter from less dense regions in the Universe by pushing it away from density minima toward the higher-density sheets, filaments, and cluster cores that make up the cosmic web. Left behind is a vast 3D network of cosmic voids that, at the present time, occupies 60–70% of the total volume of the Universe. As time proceeds, this fraction grows larger and larger. What have astronomers and cosmologists learned about the Universe in the past forty years by studying the intertwined network of cosmic voids and supercluster structure, and what can we expect to learn in the near future? These questions are the topics discussed in this final chapter.

Like all structure in the Universe, cosmic voids emerge from a distribution of small amplitude irregularities in the dark matter distribution in the early Universe. Dark matter irregularities are caused by isocurvature fluctuations meaning that irregularities in the dark matter distribution are locally compensated by opposing fluctuations in the baryons and radiation (Peebles 1993, p. 622). Furthermore, nearly the same amplitude of irregularities appears on all scales. These disturbances have their origins, most likely in or just prior to the era of inflation. They seem to have been created on all scales, even scales so large that they might extend beyond our current horizon. One fundamental aspect is that both positive and negative density irregularities are present. Hubble imagined only positive perturbations in an otherwise uniform distribution of galaxies. In the new picture, positive enhancements grow stronger over time to define the sites for galaxy and supercluster formation, while those regions with diminished initial density evolve and grow into cosmic voids, again on all scales. Tests of the standard model of cosmology (Lambda Cold Dark Matter – LCDM) are multifaceted depending simultaneously on many observed quantities including the abundance of the light elements, the universal outward "Hubble flow" of galaxies, the added outward acceleration of galaxies as first detected with Type Ia supernovae, and the tiny but detailed bumps detected in the CMB radiation. We can now add to this list of properties the observed characteristics of cosmic voids and galaxies in the large-scale structure. They all need to be considered as integral parts of the complete, optimized LCDM model. This is the basis for the statement that the discovery of cosmic voids – and the intertwined supercluster structure – contributes directly to the foundation of cosmology.

When the first N-body computer models of the Universe were built to simulate the formation of this structure, whether it was in the West with the hierarchical concepts favored by Peebles or in the East with inspiration from Zeldovich, what eventually emerged was a clear need for dark or unseen constituents in addition to the baryons we see around us. The evolution of ordinary matter cannot by itself explain the observed distribution of galaxies, superclusters, and voids and, at the same time, remain consistent with all other observed properties of the Universe. As described in Chapter 7, the early N-body simulations of structure growth – that were built by Doroshkevich, Shandarin, Klypin, and Melott – at first used neutrino dark matter with the Zeldovich concepts in an effort to explain the supercluster and void network, but when that effort failed to match the structure visible in the early 3D redshift maps, CDM was considered. The effectiveness of CDM was reaffirmed in 1985 by Davis, White, Efstathiou, and Frenk and extended further by this same group in 1987 to include supercluster structure. White (2017) highlighted in his acceptance address for the 2017 Shaw Prize the connection between computer simulations and fundamental physics. As White stated, the computer simulations showed that no known particle from the Standard Model of particle physics can account for the dark matter in such a way that the observations of cosmic voids can be explained. What this means is that White and his collaborators found that dark matter consisting of neutrinos, a conventional particle from the Standard Model, and the last hope at that time for conventional dark matter, does a poor job by itself fitting the 3D galaxy distribution, so something else, namely CDM, must be added. Of course, to arrive at this result, the N-body simulations were compared to 3D redshift survey observations of the galaxy distribution: quite specifically to the sizes of cosmic voids known at that time. So, the interplay involved all of the following disciplines: particle physics, astronomy, the physics of N-body simulations, and cosmology. This marked the first milestone in the application of cosmic voids to cosmology.

Cosmic voids have characteristics that make them a unique tool for investigations in physics and cosmology. By definition, voids arise in pristine low-density environments far from the disruptive and at times violent activity of ordinary baryons that can make a mix-master of the original matter distribution. In the realms where voids flourish, matter drifts gracefully and slowly from its point of origin. This and their diaphanous structure make cosmic voids an ideal probe to test potentially interesting extensions of the standard LCDM model. Quite specifically, tests are underway to see if the cosmological constant is sufficient to explain the accelerated expansion of the Universe or whether some form of dark energy fits better. Or perhaps the force of gravity needs to be modified on the largest scales? The dynamics of cosmic voids provides the means to carry out these tests. Finally, the properties of the rare galaxies that inhabit cosmic voids have been studied for many years by asking whether they differ from galaxies that reside in more normal environments. This makes cosmic voids a tool in the study of galaxy formation as well.

9.1   Modern Galaxy Surveys Aim for Precision Cosmology

To obtain precision results for any of these fundamental questions, astronomers need to identify and catalog very large samples of cosmic voids. The 3D border of a single void is elusive and difficult to define. This means that recipes must be developed to reliably locate voids in galaxy surveys. There are several different galaxy survey methods, each with advantages and disadvantages. The obvious first choice is to search for empty regions within the volume sampled in a galaxy redshift survey. Surveys must be large relative to the pioneering redshift surveys of the 1970s and 1980s to obtain significant samples. A deep galaxy redshift survey is a very popular choice as the starting point for current void surveys, but large allocations of telescope time are needed to obtain the galaxy spectra. Another choice is to substitute what are called "photometric redshifts" for conventional spectroscopic redshifts.[1] This alleviates the need for a majority of the telescope time, but the results can be less precise. The final choice is to rely on imaging surveys alone. Forefront work is now revealing beautiful results from deep imaging surveys by themselves.

Once identified, the 3D sizes, positions, shapes, and number density of voids (or collections of so-called "stacked voids" analyzed as an ensemble) can be tested against predictions of cosmological models. Studies of this nature have recently matured significantly, as astronomers have pushed galaxy surveys deeper into the Universe. In the early days, say before 1990, the number of voids that could be identified in a precise way was so small that statistical studies were all but useless. Therefore, in these earlier times, Blumenthal et al. (1992) suggested using extreme statistics based on the largest cosmic voids. A more recent view of this same method is given in Sahlen et al. (2016). Today, however, a revolution in survey astronomy is opening a path to identify very large void samples. Before describing how cosmic voids are used to test the physics of our Universe, I now take a small diversion to introduce three amazingly ambitious redshift survey projects. This description will not be a complete and comprehensive review of every redshift survey – there are other important surveys not discussed here – but I provide a "taste" of the work to illustrate how cosmic void research has progressed.

These new redshift surveys are certainly in the realm of "big science." They represent a change both in perspective and in research style that is nothing short of radical compared to the scientific programs conducted by those of us who did the pioneering work in the mid-1970s. We formed collaborations of two or three or maybe five scientists and used telescopes that were shared by many similarly small scientific programs. Our research expenses were modest or nonexistent. The new projects are conducted with million-dollar budgets and are run much like an industrial endeavor. Hundreds of scientists are involved, and each program has, in one form or another, dedicated observing facilities. In our original redshift surveys, we pointed the telescope toward one galaxy at a time to collect the galaxy spectra. All modern surveys use optical fibers placed in the telescope focal plane precisely aligned to the positions of their target galaxies. The fibers are all routed individually to carry the light from each galaxy to one or more spectrographs that simultaneously record hundreds of galaxy spectra. In the 1970s, we considered a 1,000-galaxy redshift survey to be very ambitious, but if a modern spectrograph can collect 500 spectra in a single exposure (500 fibers aligned to 500 galaxies), then today a survey with 500,000 spectra is comparable to the old one in terms of time at the telescope.

The first two projects in this style were the 2dF-GRS and the Sloan Digital Sky Survey (SDSS). The SDSS set as its initial goal to record with CCD detectors a major portion of the northern sky – visible from the site of its 2.5-m telescope at Apache Point Observatory in southern New Mexico – and to measure approximately 1 million redshifts (York et al. 2000). The survey has been remarkably successful. Data collection started in 1998 with the imaging portion of the survey. This initial step was finished in 2009. Images were collected and cataloged for 500 million objects recorded through five different color filters sampling galaxy images from ultraviolet to far red wavelengths. Once the images in any field of view were cataloged, the observations with a fiber-fed spectrograph began one field of view at a time. The SDSS covers 14,555 square degrees (35% of the entire sky), and the main galaxy sample has a median redshift of $z = 0.10$. For the sake of comparison, the SDSS spans an area on the sky 56 times larger and extends ~4.5 times deeper than the original Gregory and Thompson 1978 Coma/A1367 study, giving an increased sample volume of 250 times when compared with our pioneering work. The SDSS group provides to the astronomy community the survey results in what they call public Data Releases (DR), and they do this incrementally as the project matures and moves toward completion. For example, the ninth release (DR9) went public on the last day of July 2012. The following discussion of cosmic void catalogues might use words such as SDSS DR9. As of the release date of DR9, the survey had collected and analyzed spectra for 3 million objects in the main survey and an additional 800,000 new spectra in the more distant Universe, for what is called the Baryonic Oscillation Spectroscopic Survey[2] (BOSS), an extension of the SDSS that has produced its own highly significant results. The SDSS BOSS fiber optics spectrograph accepts light from 1,000 galaxies at once. Compare this with the total 238-galaxy sample of Gregory and Thompson in Coma/A1367 and the total 1,100-galaxy sample in the "Slice of the Universe" by de Lapparent, Geller, & Huchra. SDSS was started by US-based project scientists primarily at the University of Chicago, Princeton University, and Johns Hopkins University with funds from the Sloan Foundation, but it now includes additional support from Japan, the Max Planck Society in Germany, and several other partners.

The second survey briefly described here (by Abbott et al. 2018) is the ongoing Dark Energy Survey (DES). It is an international collaboration led by scientists from the Fermi National Accelerator Laboratory in Batavia, Illinois, where a huge-format CCD camera was built and then installed on the Cerro Tololo Inter-American Observatory's 4-m telescope in Chile. The DES team will survey in total 5,000 square degrees in five colors that (similar in this regard to the SDSS) span the spectrum from ultraviolet wavelengths to the far-red. Images were obtained in the five-year period 2013–18 to catalog ~300 million objects. Smaller patches of the sky are being imaged repeatedly in order to identify and then monitor thousands of supernovae. A separate Dark Energy Spectroscopic Instrument (DESI) is being placed at the prime focus of the 4-m telescope on Kitt Peak (the same spot where Gregory and I did our early photographic imaging survey of rich clusters in May 1975). DESI will survey ordinary galaxies out to a redshift of $z = 0.4$ and luminous red galaxies out to $z = 1$ with the ultimate aim of recording 35 million galaxy spectra to determine their redshifts. An additional sample of distant quasars will also be recorded spectroscopically. Using fiber optics to feed the spectrograph, DESI will record 5,000 spectra at a time. The spectroscopic survey is slated to begin in 2020.

The third survey described here is Euclid, a satellite mission organized and led by the European Space Agency. It involves an international team of more than 1,200 scientists and technicians. Euclid has a planned launch date of 2021 and a mission duration currently set at 6.25 years. It will survey the sky with an orbiting 1.2-m telescope using wavelengths from the green through the red and out to 2 microns in the infrared. Being above the Earth's atmosphere will make it easier for Euclid to detect distant galaxies whose light is redshifted towards infrared wavelengths. About 10 billion astronomical sources will be observed in a survey area of 15,000 square degrees, and spectroscopic redshifts will be determined for 50 million objects. In two additional fields on the sky, each spanning 20 square degrees, Euclid will survey 10 times deeper than the main Euclid 15,000 square degree survey. If we assume that the project adheres to its current schedule, the 6.25-year mission will be completed in 2028. Further information can be best found online under Euclid (the spacecraft).

9.2   Historical Survey Methods: How to Find Cosmic Voids

Now we return to a discussion of cosmic voids. The two earliest void catalogues were both published in 1985, one by Batuski and Burns (1985) and the second by Vettolani et al. (1985). Both studies are based on simple searches of a 3D sample volume. Batuski and Burns started by mapping the 3D distribution of Abell rich clusters within a redshift of $z < 0.13$. This means that the rich clusters in their sample had redshifts less than ~39,000 km s^{-1} and therefore distances less than 560 Mpc = 1.8 billion light-years. Even by 1985, redshifts were already available for at least a single bright galaxy in a majority of each Abell cluster. For the small fraction of clusters with no redshift, Batuski and Burns used an estimate instead. Next, they applied a percolation analysis to link, by varying the percolation distance, connected supercluster structure, and in the regions free of superclusters they identified 29 cosmic void candidates. No estimate of completeness in the void sample was provided: in fact, Batuski and Burns were clear to call them void candidates. Vettolani and his collaborators confined their analysis to a more limited volume (redshifts less than 11,000 km s^{-1} or $z = 0.04$) where in 1985 they could define a relatively complete sample of individual galaxies with redshifts. They divided their 3D sample volume into small cubes and looked for all empty cubes that were located more than a specific distance from the known galaxies. They varied this specified distance during the analysis. Adjacent empty cubes were joined together into contiguous volumes to define their cosmic void sample. In this same era, Bahcall and Soneira (1984) were also working on the 3D clustering properties of Abell clusters. Although they did not create a void catalogue like Batuski and Burns, they did produce a supercluster catalogue but then called attention to one extensive cosmic void; it rivaled the total volume of the Boötes void (Bahcall & Soneira 1982). In a manner similar to that of Batuski and Burns, this void was defined by the absence of rich galaxy clusters in a 3D volume defined by the redshifts of galaxy clusters. These were the very first small steps into a huge new area of study.

In the time frame 1985 to 2000, limited progress was made in the identification and cataloging of individual cosmic voids. In these years, the preferred galaxy redshift survey format continued to be the "slice," a format that efficiently probed structure to moderate depths into the Universe but was somewhat imperfect for those who want to catalogue and investigate the physical properties of individual cosmic voids. Edge effects were the primary obstacle, and "slice" surveys have excessive outer boundary area relative to their volume. Any cosmic void that sits near a survey border cannot be precisely characterized. Two well-known slice surveys from this era are included in the Chapter 8 timeline: the CfA2 and the LCRS. The CfA2 survey was used by Vogeley et al. (1991) to determine the Void Probability Function (VPF), that is, the probability that a randomly selected volume contains no galaxies (White 1979). Near the end of this early era Muller and his collaborators (Muller et. al. 2000) put to good use the LCRS to identify and study voids and to compare their properties to CDM models.

The most progress in this early period was made by two studies that, somewhat surprisingly, used galaxy samples located far into the southern skies. The first was completed by the University of Cape Town team of Guinevere Kauffmann and the late Anthony Fairall (Kauffmann and Fairall 1991). Fairall had spent his time compiling galaxy redshifts in the southern hemisphere (galaxies south of declination –30 degrees) and adding them to what he called the "Southern Redshift Catalogue." Kauffmann and Fairall created a program they called VOIDSEARCH that resembled the method developed by Vettolani and his collaborators in 1985, and with this program Kauffmann and Fairall identified 16 cosmic voids and 129 somewhat less-certain void candidates. The second early standout came from the Hebrew University in Jerusalem where Hagai El-Ad and Tsvi Piran (1997) developed the successful routine called "VoidFinder" and applied it to the separate Southern Sky Redshift Survey sample from Luiz Nicolai da Costa and his collaborators (El-Ad et al. 1996). They identified a sample of 12 cosmic voids.

Because VoidFinder was used repeatedly over a number of years, I describe how it works. For all galaxies in the redshift survey, the local galaxy density is calculated by searching for neighboring galaxies around each target galaxy. The local galaxy density is based on the distance from the target galaxy to its three nearest neighbors. Once calculated, this local galaxy density is used to decide if the galaxy belongs to a "wall" (local galaxy density is high) or if it is an isolated "field galaxy" (local galaxy density is low). El-Ad and Piran's protocol permits field galaxies to reside throughout all space, including the volume inside the cosmic voids.

VoidFinder then performs the actual void search by looking for spheres in the 3D data set that contain no wall galaxies. To do so, the survey volume is divided into a grid of cubes ~5 Mpc on a side. Every empty cube is used as the center of a "maximal sphere." The center point of the sphere is allowed to float within the small cubical volume, while the void radius is expanded until the sphere comes into contact with four "wall" galaxies. If the maximal sphere has a radius less than 10 Mpc (33 million light-years), it is discarded. Otherwise the maximal sphere becomes part of a preliminary catalogue. Once all maximal empty spheres have been identified, they are ordered by size and checked for overlap. If the overlap of a smaller void with a larger neighbor is greater than 10%, it joins its neighbor to form a single void that becomes part of the final void list. Each final void is identified as a continuous volume that contains no wall galaxies. Voids containing so-called field galaxies have interior densities that are not zero. Voids are characterized by their underdensity relative to the mean galaxy density. Merged voids are no longer simple spheres, so an effective radius is calculated based on the total merged volume. Only a small fraction of the voids consists of a single sphere of radius 10 Mpc. These are the basic rules, without going into the finer details.

After the year 2000 several new large redshift survey data sets became available. Astronomers at the Center for Astrophysics (CfA) published and publicly circulated the Updated Zwicky Catalogue (Falco et al. 1999) that included a total of 12,925 galaxy redshifts. In the year 2000, a second somewhat larger galaxy survey called the "PSCz survey" was published based on the Infrared Astronomy Satellite Point Source Catalogue (Saunders et al. 2000). It contained 15,411 galaxy redshifts distributed over the entire sky (but excluded the area hidden behind the Milky Way). In 2002, the 2dF Galaxy Redshift Survey was published (see the reference in the timeline in Chapter 8) with 245,591 galaxy redshifts from a 1,500-square-degree survey area. Finally, the SDSS began to release redshift survey observations with DR5.

In a 2008 paper published by Jorg Colberg entitled "The Aspen-Amsterdam Void Finder Comparison Project," Colberg et al. (2008) brought together thirteen different groups of scientists who had been working to identify cosmic voids and arranged for each group to analyze the same 3D test volume filled with simulated dark matter halos (i.e., galaxies in a simulated LCDM universe). Each group used its own previously developed void search method to identify cosmic voids in the test volume. Colberg's comparison revealed significant discrepancies in the void catalogue outcomes, clearly illustrating the complexities of precisely and consistently defining cosmic voids. This was an awkward outcome because soon it would be realized by Lee and Park (2009) that the evolution over time of cosmic void shape (i.e., their ellipticity) had the potential to reveal key information on the nature of dark energy. But the studies included in Colberg's test could hardly arrive at the same list of void candidates, let alone consistently measure their shape. Even so, Colberg's paper documents the widespread cosmic void activity being pursued in the early 2000s. Nine of the thirteen groups in Colberg's study were honing their skills on their own well-understood simulated data and had not (yet) applied their methods to real catalogue data. For any statistical study of the large-scale structure, the analysis of simulated data is a key step because it provides the only means to ensure a valid outcome when the time comes to put real data to the test.

Prominent among Colberg's thirteen participating groups was the scientific team of Michael Vogeley (b. 1964) and Fiona Hoyle (b. 1976). In 2002, they published their first joint paper on cosmic voids. As a Harvard graduate student in the early 1990s, Vogeley was trained in the Geller and Huchra research group at the Center for Astrophysics (CfA) where he participated in the analysis of the original CfA2 "Slice of the Universe" (Vogeley 1991, 1993). He eventually moved to Drexel University and established his own research group. In the year 2000, Fiona Hoyle joined Vogeley's group as a postdoctoral researcher. She had already graduated from Cambridge University in math and cosmology and had completed her PhD by working on the 2dF Galaxy Redshift Survey at Durham University. In their first joint paper, Hoyle and Vogeley (2002) used El-Ad and Piran's VoidFinder to identify nineteen cosmic voids in the Updated Zwicky Catalogue and thirty-five cosmic voids in the PSCz. In their second paper, Hoyle and Vogeley (2004) identified with VoidFinder a sample of 289 cosmic voids in the 2dF Galaxy Redshift Survey. Their results represented a significant advance. Instead of pausing to discuss these milestone contributions, I move on to another publication from the same group in 2012. The first author is Danny Pan, a 2011 PhD graduate of Drexel University and a student of Vogeley. Pan et al. (2012) reports on a catalogue of 1,054 statistically significant cosmic voids from the SDSS DR7. Once again, it was VoidFinder that they used.

A key result from the 2012 Danny Pan paper is shown here in Figure 9.1. It displays the distribution of effective void radii for their sample of 1,054 cosmic voids. The distribution is consistent with the Gregory and Thompson statement "there are large regions of space with radii $r > 20$ h^{-1} Mpc where there appear to be no galaxies whatever" as quoted from the abstract of our 1978 Coma/A1367 cosmic void discovery paper. The maximum void diameter in Figure 9.1 is 33.5 h^{-1} Mpc. This can be compared to the diameter of the Boötes void. In a follow-up to the original 1981 Boötes void discovery paper, Kirshner et al. (1987) published a more complete redshift survey that reports[3] a Boötes void radius of 31.5 h^{-1} Mpc.

Two other groups listed in Colberg's 2008 Void Finder Comparison Project were developing an entirely new method to identify cosmic voids based on the concepts of Voronoi Tessellation. Their methods were not identical but closely

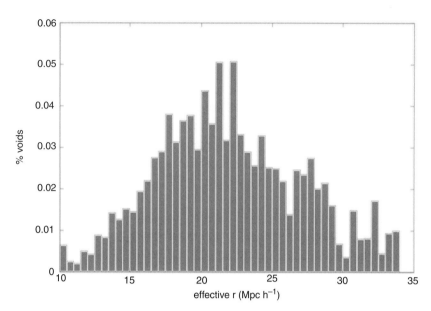

Figure 9.1 Void radius distribution. The effective void radius distribution determined by Pan et al. 2012 from the SDSS DR7. Notice that the peak in the radius distribution occurs close to the 20 h^{-1} Mpc radius found for the first two voids discovered by Gregory and Thompson (1978). The radius of the Boötes void is ~31.5 h^{-1} Mpc, close to the maximum value identified here by Pan et al. 2012. By permission of Oxford University Press: Pan, D., Vogeley, M., Hoyle, F., Choi, Y., & Park, C. (2012). *Mon. Not. Royal Astron. Soc.*, 421, pp. 926–34.

related. Neyrinck (2008) called his method ZOBOV whereas Platten, van de Weygaert, and Jones (2007) referred to theirs as the "Watershed Transform." Voronoi Tessellation is a well-known mathematical method that completely fills space with nonoverlapping close-packed cells. In this application, cells are centered on each galaxy. ZOBOV requires that the space within each cell be closest to the defining central galaxy of each Voronoi cell. If another galaxy is closer, it takes possession of that space. Of course, every cell has nearest neighbors, so cells can be zoned or grouped together with their adjacent neighboring cells. In the first step of the zoning procedure, galaxies are forced to join their lowest-density neighbor to form a set, and the process continues until every galaxy arrives at or is associated with a specific minimum density cell. The entire set of cells associated with a given minimum density is considered to be a distinct void consisting of all associated neighboring cells. At this point, the Watershed Transform method carries the analysis one step further in order to define a void hierarchy. This is done by figuratively "flooding" the volume in a manner analogous to raising the level of water in a topographical situation.

Water fills the lowest levels first (i.e., the deepest voids), but during the flooding process whenever a particular point in the volume is shared by two distinct basins, it is identified as belonging to their segmentation boundary. Once the entire volume floods, the end of the hierarchy has been reached. The fact that there are no free parameters and no restrictions as to the shape of a void gives ZOBOV (and the Watershed Transform) characteristics quite different from the void-identification routines discussed so far. One potential weakness of ZOBOV and the Watershed Transform is that many small and trivial voids are part of the final void catalogues. Those who defend ZOBOV and the Watershed Transform point out that their method provides a calculated measure of certainty for each identified void, so once a threshold is selected, the smaller catalogue members can be easily dropped from any further analysis.

Sheth and van de Weygaert (2004) discussed the evolutionary fate of both the large isolated (conventional) cosmic voids and the smaller examples that are often found by ZOBOV in somewhat denser regions. The large isolated voids tend to grow by expanding and merging with neighboring adjacent cosmic voids. The smaller ones can completely disappear during the evolutionary collapse of filaments and sheets of galaxies. This is an ongoing process that is not yet complete. Our current era (and the recent cosmological past) is a time when the large-scale structure is still forming and evolving.

Sheth and van de Weygaert also reviewed the way individual galaxies respond dynamically (i.e., how they move around or flow through space) if they happen to reside in a void. This was an old subject – even in 2004 – that was first investigated and described by Peebles (1982a) and by many others[4] in the mid-1980s. Pan et al. (2012) re-discussed the topic more recently. Each of these investigations made it clear that galaxies located within cosmic voids feel an effective gravitational push toward the nearest exterior void border. Over time, a general flow of galaxies inside a cosmic void ensues that forces a majority of the (former) void galaxies to accumulate along the outer walls of their respective voids (see Figure 9.2). This process leads to sharp ridges of galaxies often seen at the borders of voids, a phenomenon that misled de Lapparent, Geller, and Huchra to propose in 1986 that there was a need for explosive galaxy formation. Ironically, the dynamical galaxy "pile up" had already been described by Geller's thesis advisor, Peebles, four years before the 1986 publication of the Slice of the Universe redshift survey results were available.

Cosmological tests have been developed based on the observed dynamical reaction of galaxies that are fleeing the interiors of low-density cosmic voids. The easiest of these tests to describe is based on redshift space distortion (RSD).[5] Recall that redshift survey maps are not perfect renditions of 3D space. In

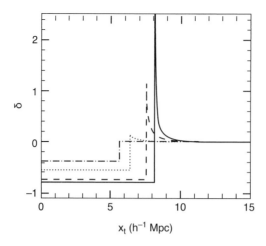

Figure 9.2 Evolution of a negative density disturbance. These results trace the evolution of a small amplitude proto-void as its interior density decreases progressively and its diameter increases over time. During this process, the void pushes galaxies that reside in the low density regions toward a "pile-up" at the outer edge. A positive value along the vertical axis (labeled "delta") represents a spatial density excess (positive value) or a spatial density deficit, that is, a cosmic void (negative value). On the horizontal axis is the void radius. The center of the void is located on the far left at $x = 0$, and the outer edge of the void evolves from $x = 6$ to $x = 8$ h^{-1} Mpc in this evolving model. Four curves reveal a time series showing how the galaxy density changes. The proto-void starts as a modest underdensity (dot-dash-dot line) that progressively grows deeper (the dotted line) as a galaxy pile-up "shoulder" begins to form at about $x = 6.5$ h^{-1} Mpc. The evolution progresses to the third step (dashed line) and ends (solid line) with a strong density enhancement represented by the sharp peak at $x = 8$ h^{-1} Mpc. The unchanging average density of the Universe is shown extending to the far right beyond a radius of 10 h^{-1} Mpc (represented by delta = 0). By permission of Oxford University Press: Sheth, R. & Van de Weygaert, R. (2004). *Mon. Not. Royal Astron. Soc.*, 350, pp. 517–38.

Chapter 5 the "Finger of God" concept was introduced when describing how the core of a rich galaxy cluster appears to be highly distended in a redshift survey map. This is an RSD. Similar RSD effects can be detected for those galaxies that reside in and around cosmic voids. Because void flow velocities are less extreme than those in rich cluster cores, a clear detection of the motion requires averaging the signal coming from many cosmic voids. The first step is to identify and catalogue a sample of voids, and then large numbers of self-similar voids are "stacked." The property of the stack yields the signal of the expected galaxy flow. Once detected, the RSD yields new cosmological information because line-of-sight redshift characteristics need to match the on-the-sky angular characteristics. Of course, these are linked to (and are sensitive to) the specific cosmology that applies to our Universe.

In the same time frame that Danny Pan and his coauthors were working on cosmic voids from the SDSS DR7 catalogue, a new precision cosmology group stepped forward to construct their own void catalogue also based on the SDSS DR7 data. The core members were Guilhem Lavaux and Benjamin Wandelt, both of whom worked at the University of Illinois in this era. This pair extended their collaboration to include David Weinberg, an experienced cosmologist from Ohio State University, and Paul Sutter, a PhD student in physics from the University of Illinois. While Danny Pan used VoidFinder for their study of the SDSS DR7 sample, this new team used ZOBOV and the Watershed Transform to produce, in 2012, a catalogue with ~1,000 cosmic voids, similar in number to those Danny Pan et al. identified with VoidFinder. Sutter et al. (2012) demonstrated their success by creating "stacked" radial profiles of voids after sorting

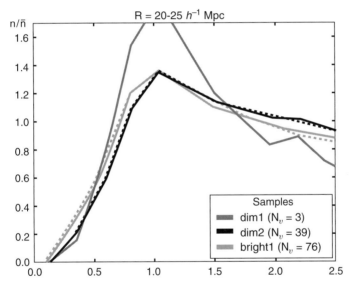

Figure 9.3 Stacked void radial profiles. The vertical axis shows galaxy counts from the SDSS DR7 Public Void Catalogue in thin spherical shells as a function (on the horizontal axis) of the radial distance from the common "stacked" void center. All voids in this particular sample have radii in the range of 20–25 h^{-1} Mpc. The three lines represent three void samples obtained from different distance ranges in the SDSS DR7 sample. Notice the peak in the galaxy counts, exactly as expected, at $R/R_o = 1$. This peak confirms in real galaxy counts, the creation of a "shoulder" or sharp edge in the distribution of galaxies at the outer rim of cosmic voids, as predicted by the simple linear theory displayed in Figure 9.2. With permission of the A.A.S.:. P. Sutter, G. Lavaux, B. Wandelt & D. Weinberg (2012). *Astrophys. J.*, 761, pp. 44–56.

them into groups of different diameters. An example is shown here in Figure 9.3. It displays in the SDSS data the expected excess or accumulation of galaxies predicted theoretically to occur at the outer wall of cosmic voids.

According to investigations published by Tully and his collaborators, our Milky Way galaxy and our nearest neighbors like the Andromeda galaxy have also partaken in this phenomenon of sweeping galaxies out of voids. In other words, these studies indicate that the Milky Way galaxy and the other galaxies around us were born in a volume that today has turned into a cosmic void – quite specifically the Local Void – and were swept into what Tully has called the "Local Sheet" situated at the outer rim of the Local Void, where a galaxy pile-up has created the planar structure shown in Figure 9.4. The analysis by Tully et al. (2008) is sufficiently sophisticated to have determined the most probable dynamical paths followed by our galaxy and those around us (Rizzi et al. 2017 and Shaya et al. 2017). Because the high-density Virgo galaxy cluster is located relatively nearby (see Figure 9.4), this cluster has also played a significant role in defining our path through space because it has gravitationally deflected our motion in its direction as we left a vast emptiness of the Local Void behind. These elegant results provide the missing link between the original discovery of cosmic voids and what showed up in the video Tully created for IAU Symposium No. 79 as described at the end of Chapter 6. It was not at all clear to me in 1979 how the large voids Gregory and I had discovered with radii ~20 h^{-1} Mpc were related to the smaller dimensions of the Local Supercluster. The Local Void itself (the solid oval in Figure 9.5) is small on this scale, but with the addition of its North and South extensions (the dashed lines in Figure 9.5), the Local Void appears to be quite respectable in size. Even so, the Local Sheet only partially covers a fraction of the outer sheath of the smaller Local Void.

The general tendency for galaxies to flow out of the lowest density regions provides a powerful diagnostic in its own right that can independently signal the presence of a cosmic void. Based on this concept, an alternate void finder was developed Lavaux and Wandelt (2010). They named their new void finder DIVA: Dynamical Void Analysis. As Lavaux and Wandelt point out, the correspondence between catalogues of empty holes in the 3D galaxy distribution – in other words, conventionally identified cosmic voids – and the structures identified by DIVA might not be a perfect one-to-one match. However, there is a clear advantage when dynamical information is available: it reveals the evolutionary state of the cosmic void. For example, voids that are contracting – being squeezed out of existence – can be easily segregated from those that are expanding (the evolutionary change expected of a "normal" void). The Zeldovich convention of describing three basic shapes of positive density enhancements – filamentary, pancake, and spheroidal – was also applied in DIVA by Lavaux and

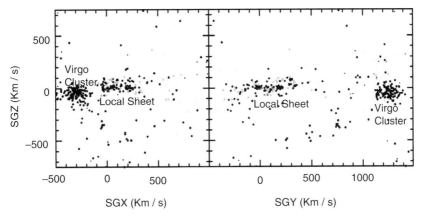

Figure 9.4 Local sheet and the Virgo Cluster: Nearby view. From these two graphs, the reader can visualize the local 3D structure in the galaxy distribution. Each point represents a single galaxy. Our Milky Way sits at coordinates (0, 0, 0). The square panel on the left shows the local galaxy distribution as viewed from one end of a rectangular volume: SGX is the Super Galactic X coordinate and SGZ is the Super Galactic Z coordinate. In this end view, the SGY distribution is collapsed into the plane of the graph. Depth information is shown in the right rectangle. Imagine cutting the entire graph out of the book and folding the right panel 90° at the line where the box meets the rectangle. Clearly, the Virgo Cluster sits behind the Local Sheet when looking into the local galaxy distribution from one end. All axes (SGX, SGY, SGZ) have units of velocity because they are derived from the "Hubble flow," but these can be approximately converted into actual distances by dividing them by the Hubble constant used in this analysis (74 km s^{-1} Mpc^{-1}). The conversion into distances is only approximate because of redshift space distortion: flow velocities distort the apparent geometry. With permission of the A.A.S.: R. Tully, E. Shaya, I. Karachentsev, H. Courtois, D. Kocevski, L. Ruzzi & A. Peel 2008, *Astrophys. J.*, vol. 676, pp. 184–205.

Wandelt to classify cosmic voids destined to take on the analogous negative density features: filamentary, pancake, and spheroidal voids. The downside of DIVA is that six coordinates are needed for each galaxy in the survey under investigation: three spatial coordinates (x, y, z) and three velocities (v_x, v_y, v_z). Because complete six-component information is not available for actual galaxy samples, at the present time DIVA can be used only to study computer-generated dark matter models of the Universe. But in this realm, DIVA is a superb tool that cannot be matched by any of the more simple-minded 3D void finders in terms of its precision.

From the start of their collaboration, the Wandelt and Lavaux team have had their sights fixed on using cosmic voids as probes to address the most significant

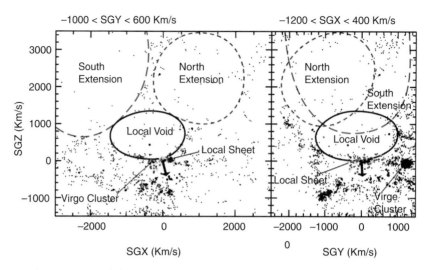

Figure 9.5 Local sheet and the Virgo Cluster: View from a greater distance.
The same two-panel visualization method used in Figure 9.4 is used again here to
show 3D structure, but in this figure a much larger volume is displayed. Each axis is
extended relative to Figure 9.4, so the displayed volume is increased by ~28×. By
doing so, the full extent of the Local Void is visible as the solid oval. Notice that our
Local Sheet sits on the rim of the Local Void. While the Local Void is somewhat small,
two likely adjacent extensions to the void are shown as dashed lines. Short dashes
outline the North Extension and long dashes outline the South Extension. The exact
sizes of both extended volumes are somewhat uncertain, because the void extensions
sit partially behind the plane of the Milky Way galaxy where local stars and dust
obscure our view of the more distant structures. The solid arrow pointing downwards
in both diagrams shows the expansion velocity vector that carries the Local Sheet
away from the central volume of the Local Void. With permission of the A.A.S.:
R. Tully, E. Shaya, I. Karachentsev, H. Courtois, D. Kocevski, L. Ruzzi & A. Peel 2008,
Astrophys. J., vol. 676, pp. 184–205.

cosmological questions. For example, in a paper published in early 2012, Lavaux
and Wandelt clearly state the unique opportunity cosmic voids provide. "If we
had a population of standard spheres scattered throughout cosmic history we
could measure the cosmological expansion directly. Absent such a population,
the next best thing is a population of objects whose average shape is spherical.
Cosmic voids are such a population and hence promising candidates for probing
the expansion history of the universe." The immediate focus of attention of
Lavaux and Wandelt (2012) was on what astronomers call the Alcock-Paczynski
(AP) Test. This test is based on the behavior over time of expanding spherically
symmetric objects (Alcock and Paczynski 1979). Their extent in the line of sight
(as measured by the redshift difference front to back) can be compared to their

apparent angular diameter on the sky. These two observable properties will be found to be equal only when the cosmological model and all adjustable parameters are correctly chosen. The potential use of cosmic voids for the AP test was first suggested by Ryden (1995). It was not until 2016 that meaningful conclusions were produced based on the AP concept as described in the following section.

9.3 Five Specific Examples of Contemporary Cosmic Void Research

A brief shift in the presentation style is made at this point, in order to place full attention on several examples of recent cosmic void research publications. Narrowing the field to these specific examples was difficult and somewhat arbitrary because there are many to choose from. Each year, since 2013, forty to fifty papers per year have been published on cosmic voids. Having already defined the essential vocabulary of redshift surveys and void cataloging, the reader can quickly grasp the significance of these new cosmological results. They speak for themselves in validating the claim that cosmic voids are a forefront contemporary tool for investigating the properties of the Universe and revealing its physical properties.

Topic: Test of General Relativity and the Matter Content of the Universe
Authors: N. Hamaus, A. Pisani, P. Sutter, G. Lavaux, S. Escoffier, B. Wandelt, and J. Weller.
Phys. Rev. Lett., 117, 091302, 2016.

"Constraints on Cosmology and Gravity from the Dynamics of Voids." In a superb analysis of the SDSS DR11 redshift survey, the authors identify cosmic voids with the void finder VIDE (basically ZOBOV, with hierarchical structure analysis from the Watershed Transform). After stacking void velocity profiles, they apply the Alcock-Paczynski (AP) test as well as a cross-correlation analysis involving galaxy positions relative to cosmic void positions. This breaks an interrelationship between parameters in the AP test and leads to an accurate determination of the mean density of matter (the combined dark matter and baryon density) in the Universe. They also find – within the errors of their measurements – no indication of any deviation from General Relativity. The analysis is most sensitive on scales of intermediate-sized cosmic voids, that is, those with radii between 30 h^{-1} and 60 h^{-1} Mpc.

Topic: Cosmic Web 3D Visualization
Authors: D. Pomarede, Y. Hoffman, H. Courtois, and R. Tully
Astrophys. J., 845, pp. 55–64, 2017

"The Cosmic V-Web." Direct distance measurements for 8,000 galaxies from Tully and his collaborators have been combined with observed galaxy redshifts to obtain "galaxy flow velocities." The flow velocities arise because galaxies attempt to flee from the low-density regions and move towards the major mass accumulations. Galaxy motions are used to determine the locations of cosmic void and supercluster structure. An animated version of these results is available in an online video through the service called Vimeo. Those who find an electronic copy of the scientific paper can click on a link in that article to view the complex topology of cosmic voids in impressive pseudo-3D images. The general concept behind this type of reconstruction was first introduced in 1989 by Bertschinger and Dekel (1989).

Topic: Neutrino Mass Determination
Authors: C. Kreisch, A. Pisani, C. Carbone, J. Liu, A. Hawken, E. Massara,
    D. Spergel, and B. Wandelt
Mon. Not. Royal Astron. Soc., 488, pp. 4413–26, 2019

"Massive Neutrinos Leave Fingerprints on Cosmic Voids." Simulated redshift survey data from CDM models are analyzed with the void finder called VIDE to demonstrate the sensitivity of void sizes to the total neutrino mass. Neutrinos cluster only on scales larger than their free-streaming length, and the free-streaming length is a function of the highest mass neutrino species: 40 h^{-1} Mpc and 130 h^{-1} Mpc structure corresponds to neutrino masses of 0.6 eV and 0.06 eV, respectively. To keep things clear, this does not hark back to neutrino-dominated dark matter models of Zeldovich, Shandarin, and Doroshkevich from the 1980s. Instead, these are ordinary CDM cosmological models where neutrinos provide only a minor fraction of the total dark matter mass. The authors anticipate that Euclid satellite data analyzed following their procedures will provide definitive results on the neutrino mass.

Topic: Tests of Modified Gravity
Authors: J. Clampitt, Y-C Cai, and Baojiu Li
Mon. Not. Royal Astron. Soc., 431, pp. 749–66, 2013

"Voids in Modified Gravity: Excursion Set Predictions." General relativity has been validated on smaller scales, like those within our Solar System, but tests on the largest scales are uncommon and difficult to accomplish. If general relativity is not applicable on scales as large as superclusters and cosmic voids, the introduction of modified gravity could replace the need for a cosmological constant and/or dark energy. These authors show how observations of cosmic voids – namely their radial profiles and the galaxy flow velocities – provide a direct and very sensitive test of models of modified gravity. The tests are designed to reveal any altered behavior on larger scales and in low-density regions of the Universe. This paper is a well-recognized pioneering work on this topic.

Topic: Measuring the Imprints of Voids and Superclusters on the CMB
Authors: Y-C Cai, M. Neyrinck, Qingqing Mao, J. Peacock, I. Szapudi, and A. Berlind
Mon. Not. Royal Astron. Soc., 466, issue 3, pp. 3364–75, 2017

"The Lensing and Temperature Imprints of Voids on the Cosmic Microwave Background." Cosmic voids are present in the line of sight toward the CMB radiation, and their presence should be detectable when the CMB photons are deflected (i.e., gravitationally lensed) by the lack of matter in the voids. These authors have tested and validated this prediction. They start with the SDSS DR12 CMASS galaxy sample and use ZOBOV to identify cosmic voids. After calculating the line-of-sight predictions, they proceed to detect the expected temperature dip in the CMB radiation as well as the signature of voids in the CMB signal. This is a significant research area and the focus of attention of multiple contemporary research groups.

This chapter closes with a discussion of three shorter topics each of which are quite separate and interesting in their own right.

9.4   Topology of the Large-Scale Structure

Jõeveer and Einasto (1978) at Tartu Observatory were the first to suggest a specific topology for the large-scale structure. As described in Chapter 5, they searched for evidence of Zeldovich supercluster pancake formation and found what seemed to them to be empty "holes" (i.e., cosmic voids) surrounded on all sides by walls of galaxies. To them, these walls were the purported Zeldovich pancakes. From this point forward, Jõeveer and Einasto proclaimed that "the Universe has a cell structure." By this, they meant that the galaxy distribution consists of a repeating close-packed honeycomb structure with dimensions – of both voids and the supercluster walls – close to 100 Mpc. Despite the lack of substantiating observational evidence for their model (cf. Oort 1983, p. 418) Jõeveer and Einasto continued to promote this concept into the early 1980s. The most widely cited of their papers on the cell structure was a review article in the publication *Nature* by Zeldovich, Einasto, and Shandarin (1982) entitled "Giant Voids in the Universe." Their speculative ideas led other scientists to consider the same "cell structure" model. For example, Takuya Matsuda and Eiji Shima (1984) of Kyoto University suggested that Einasto's model might be understood mathematically as a Voronoi tessellation. Matsuda and Shima made no check against actual observations but used a comparison with 2D computer simulations published by Melott (1983).

In the mid-1980s, Gott began a more careful and systematic study of the topology of the large-scale galaxy distribution. Gott described his work in his 2016 book entitled *The Cosmic Web*. I summarize, here, highlights of Gott's research but refer interested readers to the in-depth story told in his book. In Gott's first effort, he enlisted the help of two others: Adrian Melott, who, by this time, was building 3D models of CDM galaxy formation with realistic initial conditions (including the all-important Zeldovich Approximation), and Princeton undergraduate student Mark Dickinson (b. 1963), who was given the job of analyzing the 3D properties of the original CfA1 galaxy redshift survey data. Melott brought to the collaboration his ability to generate isocontour surfaces of galaxy density for both his models and the CfA1 redshift survey data. For example, Melott could display 3D contours that separate low-density regions from high-density regions at easily selected contrast levels. This was a necessary starting point for the topological analysis. Gott, Melott, and Dickinson (1986) demonstrated that the topology of the large-scale distribution of galaxies is sponge-like. It is not a repeating set of honeycomb cells.

Next, Gott added David Weinberg to his team, and together they introduced into the analysis what topologists call a "genus" parameter that provides a quantitative measure of the 3D topology. For example, if there had been a positive shift in the peak of the genus curve for the surface contours produced by Melott's 3D smoothing, it would have suggested a Swiss cheese or honeycomb topology, whereas if there had been a distinctly negative shift in the peak of the genus curve, it would have implied a meatball topology. Gott used the "meatball" analogy to characterize Hubble's view, where isolated clusters sit in an otherwise uniform distribution of galaxies. Gott, Weinberg, and Melott (1987) found in their analysis of the 3D CDM simulation a zero shift in the peak of the genus curve thus confirming its sponge-like topology. In a sponge-like topology, the cosmic voids and the supercluster structure both form their own continuously connected structures, and both structures are intertwined with each other in the same way that connected tunnels penetrate the body of a living sponge. One of their more remarkable conclusions relates to the evolution of the void and supercluster structure as seen in Melott's models: while the voids grow emptier and the superclusters grow denser, the spatial position of the contours – those that separate the high-density volumes from the low-density volumes – appear to change only to a minor extent from the earliest era until now.

In both the 1986 and 1987 papers, Gott and his collaborators present a convincing consistency argument that the nature of the underlying physics of the early Universe makes this topological outcome inevitable. They assert that the perturbations (i.e., the initial spatial irregularities) leading to the large-scale structure have their origin in preinflation quantum fluctuations. Positive density enhancements trigger galaxy and supercluster formation while negative perturbations trigger the formation of the network of cosmic voids. Many subsequent papers have confirmed the sponge-like topology based one a wide variety of redshift survey samples.

There is further ongoing work on the detailed topology of the cosmic web that addresses issues like the shape and origin of galaxy filaments in denser regions (Pranav et al. 2016). For the most part, these may have little to do with the topology of empty voids. By their very nature, the largest cosmic voids maintain the simplest structure, by tending, over time, to remove to their outskirts those galaxies residing in the interior volume. As mentioned above, smaller voids that occur in somewhat higher-density regions can collapse and disappear while those in lower-density regions evolve toward a more and more spherical shape (Icke 1984).

9.5   The Lemaître, Tolman, and Bondi Universe

In 1995, John Moffat (b. 1932) and his University of Toronto graduate student D. Tatarski began to discuss in a modern context a spherically symmetric but inhomogeneous model of the Universe (Moffat and Tatarski 1995). The general concept was originally suggested by Lemaître (1931b) and further investigated by Tolman (1934) and Bondi (1947), so the idea goes by the initials LTB. The basis of the model is simple. Consider a normal expanding Friedmann-Lemaître-Robertson-Walker (FLRW) Universe, but at our location, make the density lower than the cosmic average. From our perspective, at the center of the low-density pocket, the Universe would appear to be isotropic, and the only modifications relative to the simple FLRW model would occur in the radial direction. Moffat and Tatarski were inspired in 1995 to speculate about a low-density local region because ordinary cosmic voids had been discovered. Rather than considering cosmic voids with similar characteristics to those that had already been identified, they postulated a much larger under-dense volume with a radius of up to a few hundred megaparsecs (300 Mpc ~1 billion light-years) and selected the name "local void" for this hypothesized low-density region. They made calculations of the cosmological consequences based on a local density that was only 20% of the average density in the remainder of the Universe.

Recognizing possible confusion with the smaller Local Void identified by Tully and his collaborators (with a diameter ~10 times less than what Moffat and Tatarski had proposed), two University of Wisconsin astronomers Hoscheit and Barger (2018) have suggested the new nomenclature – large local void (LLV) – for underdensity features with diameters 100–1,000 Mpc. If we truly live in an LLV, the LTB model predicts two observable consequences. First, as we view the more distant Universe far beyond the LLV, it will appear to show an accelerating expansion. Second, if we measure the rate of expansion of the Universe (i.e., what is commonly called the "Hubble parameter"), we should find one value for those objects studied locally (within the LLV) and a second lower value for those objects located outside of the LLV.

When, in 1997, studies of distant type Ia supernovae showed a clear signal for an accelerated expansion of the Universe, an overwhelming majority of astronomers and cosmologists ascribed it to either dark energy or to a cosmological constant. Some attention was also given to the LTB model until a comprehensive analysis of standard cosmological observations showed that the LTB explanation cannot, by itself – and in a way consistent with the WMAP and Planck studies of the Cosmic Microwave Background (CMB) – explain the accelerated expansion of the Universe (Moss et al. 2011; Riess et al. 2011). The second observable consequence of the LTB model is a predicted shift in the Hubble parameter from a locally high value to a more distant low value. Current studies of the Hubble parameter present exactly this situation. The most recent value of the Hubble parameter measured by the Planck satellite gives $H = 67.4 \pm 0.5$ km s^{-1} Mpc^{-1} for the more distant Universe, but local measurements based on Cepheid variable stars and SN Ia give $H = 73.5 \pm 1.62$ km s^{-1} Mpc^{-1} (Kenworthy et al. 2019). Some have suggested that it may be possible that a slightly under-dense LLV centered on the Milky Way could explain a small part of this discrepancy and yet not violate the constraints defined by the Planck satellite measurements of the CMB.

As discussed in Chapter 2, observational astronomers have the job of determining the extent to which the Universe is isotropic and homogeneous. Although Hubble and Shapley had inadequate observations to address these matters in a definitive way, today such answers are nearly within astronomers' grasp. Gregory and I uncovered incontrovertible evidence for inhomogeneities in the distribution of galaxies on a scale of 100 Mpc, but on scales somewhat larger than this, say 250 to 300 Mpc, it appears that the Universe begins to approach some semblance of regularity. There are two parts to consider: isotropy and homogeneity. For the first of these factors, a recent study of the galaxy distribution based on the SDSS DR12 reports (Sarkar et al. 2019) that "the observed anisotropy diminishes with increasing length scales and nearly plateaus out beyond a length scale of 200 h^{-1} Mpc." On the other hand, the homogeneity with depth is still being investigated. Several groups have argued for a number of years that the local volume is inhomogeneous in the sense that we may live in a somewhat under-dense pocket of the Universe (Kenan et al. 2013; Whitbourn & Shanks 2016; Hoscheit and Barger 2018). The final verdict on this point is still not yet available because these studies have surveyed only small sections of the sky. A broader investigation covering a larger fraction of the sky is needed before decisive conclusions can be reached (Kenworthy et al. 2019).

The CMB radiation can also be used to study the isotropy of the Universe, and in this regard, there may be room for a few surprises. Based on nearly all measures of the CMB radiation, it conforms quite precisely to the expectations of the homogeneous and isotropic standard LCDM model. But there are a few exceptions. One involves a slight deficiency of "large-scale power" in the temperature fluctuations of the CMB. In other words, the CMB temperature distribution over the entire sky is slightly smoother than might have been expected from a random distribution of primordial irregularities. Another exception is called the "cosmic hemispheric asymmetry" meaning that in half of the sky the CMB radiation has a higher amplitude (i.e., it is brighter) by about ~14%. The hemispherical brightening has been detected by all three CMB satellites: COBE, WMAP, and Planck. Those who study the CMB find these slight anomalies to be interesting, but they are insufficient to cause alarm (Shaikh et al. 2019). One point is very clear: a major irregularity within the LLV as suggested in 1995 by Moffat and Tatarski is clearly excluded by both CMB observations and by the galaxy distribution.

9.6   Void Galaxies

As described in the first half of this chapter, occasionally one or even several isolated galaxies are found to reside inside an otherwise huge empty cosmic void. Soon after voids were discovered, curiosity about these so-called "void galaxies" drove astronomers to identify as many examples as possible and to study them as a class. The basic idea was to identify unique characteristics of void galaxies that might set them apart from more normal galaxies and thereby reveal information about galaxy formation and galaxy evolution. For example, Bothun et al. (1985, 1986) asked whether cosmic voids are preferentially filled with low-surface-brightness galaxies. The answer to that question is "no." However, the very brightest galaxies are not found in voids. Other groups asked whether void galaxies show a higher fraction of emission lines in their spectra, a characteristic that would indicate more active star formation for void galaxies (Sanduleak & Pesch 1987; Moody et al. 1987; Weistrop et al. 1988). The effort led by Moody involved both Kirshner and Gregory. To answer these questions requires a control sample against which to make the comparison, and that control sample is generally tied to galaxies like our own Milky Way and our near neighbor, the Andromeda galaxy. Often, the control sample is enlarged to include all galaxies except those that are members of dense and richly populated regions where galaxy clusters – like Virgo and Coma – reside. (Cluster galaxies have their own evolutionary characteristics, caused by the high density of dark matter, hot intra-cluster gas and galaxies in their immediate vicinity, that sets them apart from control samples that reside in lower-density regions.) Although studies of void galaxies first appeared in the mid-1980s, more than thirty years later the same quest is being pursued (Beygu et al. 2017). Sophisticated theoretical concepts have been discussed in an attempt to predict what differences might be expected (e.g., Aragon-Calvo and Szalay 2013).

When searching for differences between void galaxies and the ordinary galaxy population, one precaution must be highlighted. The environmental contrast between void galaxies and nonvoid galaxies appears to be drastic today, but this was not always the case. The dramatic features we see today in the large-scale structure, with nearly empty cosmic voids, are evolving characteristics that appear most prominently in our current epoch. In earlier epochs, galaxies may have shared more or less similar environments. This possibility is evident in the computer visualization created by Edward Shaya of the University of Maryland, who works in close collaboration with Tully and the other members of Tully's group of researchers (Shaya et al. 2017). By using what is called the "least action principle," Shaya identified the most likely paths followed by each galaxy in our neighborhood of the Universe, from their birth until the present. This includes the reconstructed paths for the Milky Way and our Andromeda neighbor. These calculations demonstrate that our galaxy was formed within and then exited the volume now occupied by the Local Void. No doubt, carried along with the galaxies in this gentle but steady dynamical flow was the reservoir of gas that surrounds each galaxy. This cosmic gas is known to cycle in and out of the host galaxy as violent events within each galaxy – events like supernovae – eject gas from the galaxy, and then the ejected gas is recycled by turning around and falling back towards active star-forming regions. Because those galaxies that are left behind in a cosmic void are left undisturbed, they may have maintained a more continuous history of recycling. This might set void galaxies apart from galaxies that find themselves in more crowded environments. This scenario is more sophisticated than the concepts discussed in the 1980s when most astronomers assumed that cosmic voids have been empty for eons and that, perhaps, cosmic voids might be filled with a strange collection of dwarf or low-surface-brightness galaxies. In fact, models of galaxy formation in the CDM paradigm, with the added concept of the halo occupation distribution (i.e., how ordinary matter is matched to the CDM halos) have now diffused tensions (Tinker and Conroy 2009) that were originally suggested by Peebles (2001) when he asked: Why are cosmic voids not filled with dwarf galaxies? Cumulatively over the past fifteen years, void galaxies studied from the SDSS sample have been shown to be somewhat smaller in diameter and to be dominated by star-forming spiral and irregular galaxies when they are compared to their analogs that reside outside of voids (Beygu et al. 2017; van de Weygaert et al. 2011; Grogin and Geller 1999; Rojas et al. 2004; Ricciardelli et al. 2014; Pisani et al 2019). The last of these references is especially insightful and forward-looking, and was helpful in preparing this summary.

Cosmic voids have already earned their place in the pantheon of the most significant astronomical objects and are providing new avenues to explore basic physics and cosmological phenomena. Thanks to the insights of Shaya, Tully, and their collaborators, we can say that our own Milky Way galaxy and our nearest neighbor, Andromeda, had their origins deep within the Local Void, and since that time we have been gently pushed away – by a gravitationally induced dynamical flow – from that now-vacated volume to our new vantage point in the outskirts of the Local Supercluster. From here, astronomers survey the visible Universe and address questions about the fundamental nature of cosmology.

Appendix A
KPNO Observing Proposal

SPECIAL REVIEW_____, _____   CODE NO._____

KITT PEAK NATIONAL OBSERVATORY OBSERVING REQUEST

Date Received_____

- -

NAME_Stephen A. Gregory   and Laird A. Thompson_____

INSTITUTION_SUNY Oswego (SAG)   and Kitt Peak (LAT)_____

PROGRAM____Spectroscopy of Galaxies Between the Coma and A1367 Clusters____

TELESCOPE _2.1 meter_____   DATES__April or May new moon_____

DAYS_none_____ NIGHTS__4 dark_____ (DARK GREY BRIGHT)

EQUIPMENT REQUESTED____Cassegrain image tube spectrograph (gctd)_____

Recent KPNO telescope allocations: May 1975 three nights on 4 meter.

Recent publications based on KPNO support: Reduction of above data is
   now under way. No publications yet.

210

OBSERVING TIME REQUEST FORM

Kitt Peak National Observatory

     Please prepare requests as completely as possible in order
to prevent delays and excessive correspondence, and submit prior
to 30 April for August-December scheduling consideration and
prior to 31 August for the scheduling interval January-July.
See Policies for Visitors (SECTION 1) in the KPNO Facilities
Notebook for further details.

1.   Name: Stephen A. Gregory   and   Laird A. Thompson

2.   Address and telephone number:
     (SAG)   Dept. of Earth Sciences   (LAT)   Kitt Peak National Observatory
             SUNY College at Oswego
             Oswego, New York 13126
             (315) 341-4251   (602) 327-5511
3.   Title of proposed program:
             Spectroscopy of Galaxies Between the Coma and A1367 Clusters

4.   Will you apply for partial travel support according to KPNO
     policies? Gregory would like to apply for travel support.

5.   Scientific justification. State concisely the background,
     purpose, and significance of the program and what you hope
     to accomplish. Continue on supplementary pages if necessary.

                 see attached sheets

5. Scientific Justification

We would like to determine redshifts for a moderately large sample of
galaxies which appear to fall between the two rich clusters A 1656 (Coma)
and A 1367. This study is interesting for a number of reasons, but before
outlining these reasons it is important to point out the uniqueness of this
particular intercluster region. Hauser and Peebles (1974) analyzed the
spatial distribution of rich Abell clusters and concluded that a significant
portion of clusters occur in pairs with separations $\lesssim$ 80 Mpc/h (h - Hubble
constant/100 km s^{-1} Mpc^{-1}). Because Coma and A 1367 both have nearly the
same mean redshift (z = 0.022), the angular distance between the two cluster
centers indicates that they are separated in space by only ~ 40 Mpc/h. Consequently Coma and A 1367 are probably closely related to one another, and
in this study we hope to obtain data which will help to show just what this
relationship might be. Specifically, we will use the data to answer the
following questions:

(1) What is the galaxy number density (and hence mass density) in the
region between the two clusters?

(2) Are the intercluster galaxies which have redshifts similar to the
two rich cluster cores distributed between the clusters in a way which is
consistent with the conventional view that clusters formed by the infall of
galaxies toward regions of slight density enhancement?

(3) Do the Coma cluster ($<v_o>$ = 6950 km s^{-1}) and A 1367 ($<v_o>$ = 6400
km s^{-1}) merge smoothly together over the 19° region, or is there a discon-
tinuous jump in the redshift distribution?

(4) There are two small clusters listed in Zwicky's catalogue which
fall within the area between Coma and A 1367 (#16 in Field 128 and #2 in
Field 98). Do these two separate condensations lie directly between the two
rich clusters, or are they foreground or background condensations? What
are the velocity dispersions of these two groups?

(5) What is the velocity dispersion for those galaxies which are located between the clusters, and is this dispersion constant throughout the
intercluster region?

(6) Are there many foreground or background groups in this region
similar to those found in the direction of the Coma Cluster (Tifft and
Gregory 1975)?

In carrying out this survey we intend to coordinate closely with other
investigators who have worked in this same area of the sky. Three other studies
are of particular interest. First, redshifts for the cluster A 1367 have
been determined by Tarenghi and Tifft (1975). Second, the entire Coma cluster

5. (con't.)

region was surveyed out to a radial distance of 6°, and redshifts are known
for many galaxies in this area (Tifft and Gregory 1975). And third, Rood
and Chincarini are studying a Zwicky cluster which is somewhat to the north
of our proposed area. We will select our sample to avoid any overlap with
the above-mentioned studies.

References

Hauser, M. G. and Peebles, P. J. E. 1973, Ap. J. 185, 757.
Tarenghi, M. and Tifft, W. G. 1975, Ap. J. (Lett.) 198, L7.
Tifft, W. G. and Gregory, S. A. 1975, preprint.

6. Details of the program

     The galaxies which we intend to observe will all be chosen from the
intercluster area 4° wide and 19° long. The galaxies' positions will range
in R. A. from 11^h50^m through 12^h45^m and in declination from +20° through +30°.
We intend to choose galaxies from the Zwicky catalogue with the highest
priority being given to galaxies in the fainter magnitude range $15.0 \leq m_p \leq 15.7$.
The lowest priority will be given to the very brightest galaxies ($m_p < 14.1$)
since these are almost always members of the Local Supercluster. Furthermore,
priorities will be chosen to ensure that the two Zwicky clusters lying between
Coma and A 1367 are well observed.

OBSERVING TIME REQUEST FORM, cont.

6. Details of the proposed program, including the specific objects
   or group of objects, positions or range of positions, magnitudes
   and other properties defining the observational program.

                    see attached

7. Telescopes and auxiliaries required. Detail the instrument
   required for each type of observation, other auxiliary instruments required, and expendable supplies such as photographic
   plates, chemicals, bottled gases, and cryogenics:

            We need the 2.1 meter telescope with the cassegrain image tube
   spectrograph. Less than one box of 8x10 inch photographic plates can be
   cut to size. We will use the standard emulsion for this instrument (IIa-D ?).
   One batch of developing chemicals should be sufficient.

8. Estimated number of nights and lunar phase required. Specific
   dates and alternates may be stated if required.
            Four dark nights are needed. Any moonlight overwhelms the light
   from these faint galaxies. If some moonlight must be present, please let
   it be before dawn rather that at sunset. Gregory cannot observe until
   late March, so the April or May new moons are best.

OBSERVING TIME REQUEST FORM, cont.

9. Details of any proposed use of facilities at the Tucson
   headquarters.

        none anticipated

10. If the facilities of other observatories are being used or
    applied for in connection with the project, give details.

Graduate Students Only: If the investigator is a graduate student,
the application must be endorsed by a faculty advisor who will
certify that the student is in good academic standing, that the
proposal is an accepted part of dissertation research, and that
the student has the training and ability to use the instruments for
the proposed work.

Appendix B
Gregory & Thompson (1978) Reprint

THE ASTROPHYSICAL JOURNAL, **222**:784–799, 1978 June 15

THE COMA/A1367 SUPERCLUSTER AND ITS ENVIRONS

Stephen A. Gregory [*]

Department of Earth Sciences, State University of New York College at Oswego; and
Physics Department, Bowling Green State University

AND

Laird A. Thompson

Kitt Peak National Observatory; [†] and Department of Physics and Astronomy, University of Nebraska
Received 1977 September 7; accepted 1977 December 21

ABSTRACT

The three-dimensional galaxy distribution in the region of space surrounding the two rich clusters Coma and A1367 is analyzed by using a nearly complete redshift sample of 238 galaxies with $m_p < 15.0$ in a 260 degree^2 region of the sky; 44 of these redshifts are reported here for the first time. We find that the two clusters are enveloped in a common supercluster which also contains four groups and a population of isolated galaxies. The least dense portions of the Coma/A1367 supercluster have a density which is approximately 6 times that of the Local Supercluster in the regions of our own Galaxy. In front of the Coma/A1367 supercluster we find eight distinct groups or clouds but no evidence for a significant number of isolated "field" galaxies. In addition, there are large regions of space with radii $r > 20\,h^{-1}$ Mpc where there appear to be no galaxies whatever. Since tidal disruption is probably responsible for the isolated component of supercluster galaxies, the observations suggest that all galaxies are (or once were) members of groups or clusters. A number of related topics with more general significance are also discussed. (1) The size-to-separation ratio for foreground groups indicates that the epoch of group formation is $z \lesssim 9$. (2) There is a general correlation between the volume mass density of a galaxy system and the morphologies of the component galaxies. (3) Finally, we speculate that all clusters of richness class $\mathfrak{i} \geq 2$ are located in superclusters.

Subject headings: galaxies: clusters of — galaxies: redshifts

I. INTRODUCTION

Galaxies appear to be organized on the largest scale in extensive, second-order clusters called *superclusters.* The early controversy over this matter (see Abell 1975) has been settled by Peebles and his collaborators (see Peebles 1974), who have studied the projected distribution of galaxies on the sky. More detailed information on superclustering is available if a third spatial dimension can be added directly. For example, Rood (1976) studied the three-dimensional distribution of nearby clusters by assuming that the average redshift of the observed members of a cluster is an accurate indicator of the cluster's distance, and his findings are consistent with earlier results. Our purpose in this paper is to use a redshift-based technique for studying the three-dimensional distribution of a large number of galaxies in the region of the sky defined by $11^h5 \leqslant \alpha \leqslant 13^h3$, $19° \leqslant \delta \leqslant 32°$. Within this surveyed region lie the two rich clusters Coma and A1367, and the results presented below will show that they are embedded in a common, very large supercluster.

The existence of the Coma/A1367 supercluster per se was not suggested until recently (Tifft and Gregory 1976; Chincarini and Rood 1976). We note that Abell

(1961) thought that there might be a supercluster containing six clusters centered at 11^h45^m, $+29°5$ (1950) which had a diameter of $45\,h^{-1}$ Mpc and would therefore include Coma and A1367. However, this suggestion was made before redshifts were available for any of the clusters except Coma. Two of the members, A1185 and A1213, are now known to have significantly higher redshifts than Coma and A1367 (Noonan 1973). The present study is the first that can actually demonstrate that Coma and A1367 form a unified system. The reason that the Supercluster[1] was not recognized long ago is that the two major cluster centers are widely separated. Hauser and Peebles (1973) found that rich cluster pairs are correlated if their separation is less than $\sim 20\,h^{-1}$ Mpc, where $h \equiv H_0/100$ km s^{-1} Mpc^{-1}. For the case of Coma and A1367 the separation is $21\,h^{-1}$ Mpc. This wide physical separation coupled with the Supercluster's proximity to our own Galaxy (the distance to the Supercluster is $\sim 70\,h^{-1}$ Mpc) make the angular

[*] Visiting Astronomer at Kitt Peak National Observatory.

[†] Operated by AURA, Inc., under NSF contract AST 74-04129.

[1] Hereafter we will use the proper noun "Supercluster" when referring to the Coma/A1367 supercluster if no confusion results. Following current usage, the general term "supercluster" will be used when referring to groups of galaxy clusters. At times in the past (cf. Shane 1975), these groups of clusters have been called "clouds," a term which is now reserved for loose aggregates of galaxies such as the Coma I cloud.

separation $\sim 20°$, which is too large to be visually impressive at, say, the scale of the Sky Survey.

The very large sizes of superclusters offer compelling reason for studying them in detail. Internal mixing cannot be well advanced, so the present distributions of galaxy luminosity and morphology must reflect to some extent the properties of the primordial supercluster material. We will show in the following analysis that there are three observationally distinct populations of galaxies within the Supercluster. These are (1) galaxies located in the two rich cluster cores, (2) galaxies located in intermediate- or low-mass clusters, and (3) a nearly homogeneously distributed population of isolated galaxies.

An important additional benefit of our magnitude-limited survey is that it samples foreground galaxies in addition to those in the more distant Supercluster. This adds a fourth distinct population to the three found in the Supercluster. The foreground galaxies are found in low-mass clusters; they are not distributed in a homogeneous "field." The addition of these foreground systems to our sample of clusters enables us to examine intrinsic properties of clusters over a range of $\sim 10^2$ in mass.

II. THE REDSHIFT SURVEY

The purpose of our new redshift program is to provide a sample of galaxies complete to a limiting magnitude of $m_p < 15.0$ in the area of the sky encompassing the two rich clusters Coma (A1656) and A1367. Previous redshift observations are summarized by Tifft and Gregory (1976) for galaxies with $m_p < 15.0$ and $r < 6°$ from the Coma cluster center, by Chincarini and Rood (1976) for galaxies with $m_p \leq 15.0$ in a region directly west of Coma, and by both Tifft and Tarenghi (1975) and Dickens and Moss (1976) for galaxies near the center of A1367.

Although the above mentioned surveys cover a substantial region of the sky, they leave a large gap along a line connecting Coma and A1367 in which very few galaxies have been studied. For our new observations we tried to observe each apparently isolated galaxy in this gap for which no previous redshift determination existed. However, because of observing time limitations and two aborted attempts on very low-surface-brightness objects, five of the galaxies lying within $r \leqslant 3°$ of the line joining the two clusters still do not have measured redshifts. In addition to the apparently isolated galaxies, there were five obvious groups for which no distance information existed. We tried to be as complete as possible in obtaining redshifts in the two groups nearest to A1367 so that their relationship to the rich cluster might be clarified. For the remaining groups we assumed that the redshifts of the one or two brightest galaxies were representative of the whole.

We obtained the new spectra during the nights 1976 April 27–28 to April 30–May 1 with the Kitt Peak 2.1 m telescope equipped with the white CIT spectrograph and a HeNeAr comparison source. The 300 line mm^{-1} grating gave a dispersion of 240 Å mm^{-1} in the blue, and the spectra were recorded on baked

IIIa-J spectroscopic plates. The spectrograms were measured with either a Mann comparator or a Grant measuring engine with scanning display. No significant difference was found between the results obtained for selected spectrograms measured on both instruments. The [O I] $\lambda = 5577.35$ night-sky line was measured and used to correct for systematic spectrograph or measuring errors; the average correction was $+6 \pm 28$ km s^{-1}. We used the effective wavelengths given by Sandage (1975), and we corrected for 300 km s^{-1} galactic rotation.

In Table 1 we list the new redshifts along with other pertinent data. Column (1) contains the identification number from Zwicky and Herzog (1963, hereafter CGCG). The first three digits give the CGCG field number, and the last three digits give the sequential galaxy number. Columns (2), (3), (4), and (5) give the NGC or IC numbers, the 1950 epoch right ascension and declination, and the photographic magnitude m_p. In column (6) we list the new redshift determinations. Typical uncertainties for redshifts of galaxies with $13 \leq m_p \leq 15$ at this dispersion are ± 100 km s^{-1}; parentheses in column (6) denote one particularly uncertain redshift and one which was estimated. Column (7) lists those spectral lines which were measured with nonzero weight. The lines $\lambda = 3727$ and $\lambda = 5007$ were, of course, always in emission. Hydrogen Balmer lines were found in absorption unless denoted by (em). Because of possible blending with Hϵ, the Ca I H-line was not used if any Balmer lines of shorter wavelength than Hβ were visible.

When the 44 new redshifts are combined with those in the literature, we have a total sample of 238 galaxies with $m_p < 15.0$ in the surveyed area.

III. RESULTS

a) Overview of the Supercluster Area

Although we are most interested in studying the galaxies located between and in the immediate vicinity of the Coma and A1367 clusters, it is of importance to investigate the total possible extent of the Supercluster. Consequently, we show in Figure 1 an isophotal diagram for the region of the sky which contains the Supercluster; the surface area of this diagram is ~ 10 times that of our redshift survey. The contours represent the luminosity distribution of galaxies at surface-brightness intervals of 0.5 mag. The diagram was constructed by averaging at 1° centers the total luminosity of those galaxies which fall within a radial distance of 1° from each averaging center. All galaxy magnitudes and positions were taken from a magnetic tape version of CGCG, and the averaging process was carried out by computer at the Kitt Peak National Observatory. Because the surface-brightness contours are by definition luminosity-weighted, background contamination is insignificant. Foreground contamination can be substantial, and two techniques were used to reduce this problem. First, no galaxy was included in the luminosity average if it had an apparent magnitude brighter than the first brightest galaxy in either A1367 or Coma (NGC 4889 has $m_p = 13.0$). Second,

TABLE 1

New Redshifts

Zwicky No.	Name	α(1950)	δ(1950)	m_p	V_0	Spectral Lines Used	Notes
097030	N 3768	11^h34^m6	$+18°07'$	13.7	3288	K,H,5175	
097051	N 3801	11 37.7	+18 00	13.3	3356	K,4384,5269	
097161	N 3919	11 48.0	+20 17	14.5	6052	H,G,4384,5175	
127022	N 3798	11 37.6	+24 59	13.9	3459	K,Hδ	
127027	N 3812	11 38.5	+25 07	13.9	3529	K,H	
127050		11 43.4	+21 19	14.8	(6736)	H	1
127063	N 3910	11 47.4	+21 38	14.4	7768	K,G,5175	
127064	N 3920	11 47.5	+25 13	14.1	3547	3727,K,Hδ,Hγ,5007	
127076N		11 48.9	+22 19	--	8514	H,G	2
127076S	N 3926	11 48.9	+22 19	14.1	7464	K,H	2
127080	N 3929	11 49.1	+21 17	14.5	7041	K,H,G	
127086		11 50.0	+23 53	14.8	6793	K,H,G	
127088	N 3937	11 50.1	+20 55	14.0	6554	K,G,5175	
127089	N 3940	11 50.1	+21 17	14.3	6341	K,G,4384	
127090	N 3943	11 50.3	+20 46	14.7	6559	H,G	
127098	N 3954	11 51.1	+21 10	14.4	6800	K,H,G	
127099	N 3951	11 51.1	+23 40	14.5	6430	K,H	
127100		11 51.4	+20 52	14.9	6840	K,H,G	
127110	N 3987	11 54.7	+25 29	14.4	4508	K,H	
127115	N 4003	11 55.4	+23 24	14.8	6438	K,H	
127120	N 4005	11 55.6	+25 24	14.1	4333	K,H,Hδ,4226,G,5175	
128003		12 00.9	+22 30	14.6	6476	3727,K,H,4384,Hβ (em)	
128005	N 4061	12 01.5	+20 30	14.4	7072	K,H,5175	
128007	N 4065	12 01.6	+20 30	14.0	6216	K,H,5269	
128008	N 4066	12 01.6	+20 38	14.4	7294	H,G	
128009	N 4070	12 01.6	+20 42	14.3	7143	K,H,5175	
128017	N 4084	12 02.7	+21 30	14.9	6621	K,H,G,5269	
128020	N 4089	12 03.0	+20 50	14.9	6905	K,H,G	
128023	N 4092	12 03.3	+20 46	14.4	6737	K,H	
128025	N 4095	12 03.4	+20 51	14.6	7057	K,H,4226,G,5269	
128026	N 4098	12^h03^m5	$+20°53'$	14.5	7280	K,Hδ,Hβ	
128027	N 4101	12 03.6	+25 50	14.7	6089	K,H,G	
128034		12 05.5	+25 31	14.4	(6700)		3
128054		12 10.8	+21 55	14.6	7220	K,H,5175	
128060	N 4204	12 12.7	+20 56	14.3	690	3727	4
128065	N 4213	12 13.1	+24 15	14.3	6986	K,H	
128077	I 780	12 17.4	+26 03	14.5	6779	K,H	
128078	I 3171	12 17.9	+25 51	14.8	6935	K,H	
128089	I 791	12 24.5	+22 55	14.2	6735	K,H	
129002	N 4455	12 26.2	+23 06	13.0	588	3727	4
129012	I 3581	12 34.1	+24 42	14.9	6972	K,H	
129020	I 3692	12 40.4	+21 16	14.8	6412	K,H,G	
161042		13 21.4	+31 50	13.9	4723	K,H,G	
161056	N 5157	13 25.0	+32 17	14.4	7230	K,H,5175	

NOTES: 1. Redshift is more uncertain than usual, since only one line had non-zero weight in the determination.

2. Northern component is fainter; both appear imbedded in the same halo.

3. One side of the comparison spectrum was ruined: V_0 is an estimate.

4. Poor quality spectrum; however, λ3727 was easily measured. Accuracy of V_0 is not as poor as if only one absorption line had been measured.

two separate contour diagrams were produced, the first for $m_p \geq 13.0$ and the other for $m_p < 13.0$. The second version shows mainly local galaxies, so it was possible to identify on the first version those regions that are strongly contaminated by faint galaxies in nearby groups and clusters. Figure 1 shows only the first version (i.e., $m_p \geq 13.0$), but those features in Figure 1 that have corresponding contours in the second version are shown in dashed lines. For instance, the faint galaxies in the extensive Virgo cluster complex dominate the region centered at 12^h30^m, $+13°$, but since the contours are shown as dashed lines they are easy to ignore.

Some interesting features are obvious. First, the Coma cluster has an elliptical shape with ellipticity approximately equal to 0.5; estimates of the ellipticity have been published elsewhere (Rood et al. 1972; Thompson and Gregory 1978; Schipper and King 1978). There also seems to be a difference between the regions just exterior to the two rich clusters. A1367 is surrounded by a tenuous group of galaxy clouds, whereas the region around Coma is nearly devoid of structure. Only one concentration is seen near Coma. This is the NGC 5056 group, which may not be part of the Supercluster. The contour diagram is somewhat deceptive because the faintest illustrated isophote is

No. 3, 1978   COMA/A1367 SUPERCLUSTER

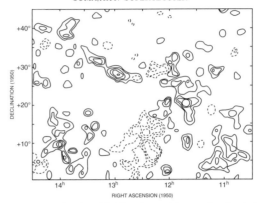

Fig. 1.—Luminosity contour diagram of the Coma/A1367 Supercluster and its surroundings for galaxies in the Zwicky catalog with $m_p \geq 13.0$. The contours are spaced at 0.5 mag intervals. The brightest contour level is 29.5 mag arcsec^{-2}, and the faintest is 31.5 mag arcsec^{-2}. *Dashed lines*, regions that are probably in the foreground. The NGC 5416 cluster ($\alpha \approx 14^h$, $\delta \approx 10°$) has a luminosity similar to that of the Coma ($\alpha \approx 13^h$, $\delta \approx 28°$) and A1367 ($\alpha \approx 11^h40^m$, $\delta \approx 20°$) clusters.

not at the limiting luminosity of CGCG but at the level beyond which the background becomes quite noisy. We will show in the following section, using the complete redshift sample to eliminate foreground and background confusion, that the region around the Coma cluster is not entirely empty but contains a widely dispersed but significant population of Supercluster galaxies.

An additional feature of Figure 1 which has potential importance is the concentration of galaxies at $\alpha = 14^h$, $\delta = +10°$. Although this concentration is nearly as prominent as A1367, it has escaped notice (for example, it has no Abell number). We will refer to this system as the NGC 5416 cluster. No redshifts are available for any of the member galaxies, but a comparison of apparent luminosity functions shows that the NGC 5416 cluster may well lie at the same distance as Coma and A1367 and may therefore indicate that the Supercluster is larger than the boundaries of our redshift survey. The NGC 5416 cluster is surrounded by an extensive system of galaxy clouds and hence is very similar to A1367.

b) Coma/A1367 Supercluster: The Interconnection

There is no widely accepted definition of the term "supercluster." At present, observational studies of superclusters use largely subjective criteria which are based on the distance and angular proximity of cluster centers. In the case of two clusters which are as widely separated as Coma and A1367, we believe that, in order to show that they are members of the same supercluster, it is both necessary and sufficient to demonstrate the presence of a population of galaxies

linking the two clusters which is itself a region of significantly enhanced density. For ease of discussion, the area between Coma and A1367 will be called the intercluster region (ICR), and the galaxies in this region will be referred to as ICR galaxies. We present a list of these galaxies in § IIIe; here we will show only that the density of the ICR is indeed significantly large.

The intercluster region is easily recognized in Figure 2a. The diagram shows the projected three-dimensional distribution of our sample galaxies. Since Coma and A1367 are separated primarily in the east-west direction, we have plotted the positions in the coordinate system right ascension versus redshift. By making the width of the R.A. axis a linear function of redshift, we have removed the major distance-dependent distortion. If the R.A. axis were of uniform width over the entire redshift range, a hypothetically spherical nearby group would appear to be unrealistically elongated in right ascension. According to the Hubble relation, redshift is proportional to distance for galaxies at rest with respect to their local comoving coordinates. However, in the relaxed cores of massive clusters galaxies have large kinetic energies, and the extreme redshifts are not indicative of distance effects. These clusters appear as very elongated structures which point toward the origin of the wedge diagram.

Two features of Figure 2a are significant. First, the clumpiness of the foreground distribution is clear. There are several groups of galaxies, and it is important to note there are large regions which are devoid of galaxies. (In fact, we note that no galaxies with $m_p < 15.0$ are found within 20 Mpc of the near side of the intercluster region.) The second point is that

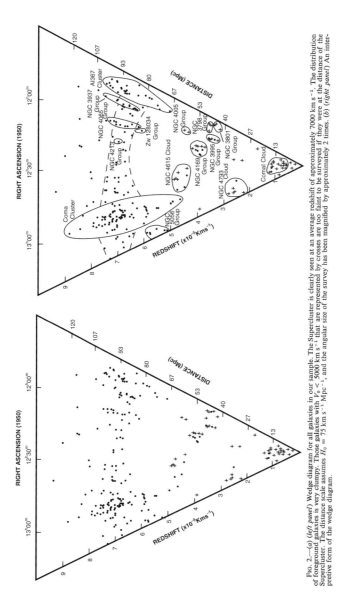

Fig. 2.—(a) (*left panel*) Wedge diagram for all galaxies in our sample. The Supercluster is clearly seen at an average redshift of approximately 7000 km s⁻¹. The distribution of foreground galaxies is very clumpy. Those galaxies with $V_0 < 5000$ km s⁻¹ that are represented by crosses are too faint to be surveyed if they were at the distance of the Supercluster. The distance scale assumes $H_0 = 75$ km s⁻¹ Mpc⁻¹, and the angular size of the survey has been magnified by approximately 2 times. (*b*) (*right panel*) An interpretive form of the wedge diagram.

Figure 2a gives a misleading visual impression of the density of the foreground. Since our survey is limited by a fixed apparent magnitude, nearby groups are studied to intrinsically fainter limits than the Supercluster. We correct for this effect by using two symbols for those galaxies with $V_0 < 5000$ km s^{-1}. Galaxies represented by large filled circles have absolute magnitudes M_p which satisfy the condition $M_p \leq M_l$, where M_l is the limiting absolute magnitude of galaxies studied in the Supercluster. With $H_0 = 75$ km s^{-1} Mpc^{-1} (assumed throughout unless explicitly stated otherwise) and a limiting apparent magnitude of $m_p = 14.9$, we find $M_l = -19.9$. Those foreground galaxies that are represented by crosses have $M_p > M_l$ and would be too faint to be included in our survey if they were members of the Supercluster. Figure 2b is an interpretive form of Figure 2a. Borders for all the galaxy systems have been added, and they will be discussed below.

We will now use two comparisons to show that the ICR has a significantly enhanced density. First, we calculate the density contrast $\rho_{\rm ICR}/\rho_f$ between the intercluster region and the foreground (f) because ρ_f is characteristic of the mean density of the universe outside of rich clusters. We take the ICR to be 6.4 h^{-1} Mpc thick (see § IIIe). The 69 ICR galaxies are distributed over 167 degrees^2 of the sky, so they occupy a volume of $1.6 \times 10^3 \, h^{-3}$ Mpc^3, giving a number density $n_{\rm ICR} = 4.4 \times 10^{-2} \, h^3$ Mpc^{-3}. In the foreground there are 13 galaxies with $M_p \leq M_l$, and they occupy the volume of space out to a redshift of $\sim$5000 km s^{-1}, corresponding to $3.3 \times 10^3 \, h^{-3}$ Mpc^3. The number density of foreground galaxies is then $n_f = 3.9 \times 10^{-3} \, h^3$ Mpc^{-3}. By assuming that the M/L ratios and luminosity functions of the two populations are identical, we find $\rho_{\rm ICR}/\rho_f = 11$.

For the second test we will compare the density of the ICR to that of the Local Supercluster (LSC) in our own vicinity. We are located at a distance from the center of the Local Supercluster that corresponds to the midpoint between the two rich clusters in the Coma/A1367 Supercluster. Jones (1976) finds that the local density of galaxies with $M_B < -18$ is $n_0 \approx 9 \times 10^{-3}$ Mpc^{-3}. We convert this density to the limiting magnitude of our survey by using the luminosity function given by Abell (1975); this gives $n_{\rm LSC} \approx 3 \times 10^{-3}$ Mpc^{-3}. Once again we assume constant M/L ratios and find $\rho_{\rm ICR}/\rho_{\rm LSC} = 15 \, h^3$. In his calculations, Jones used a value of $h = 1$, appropriate to the local vicinity of space (see van den Bergh 1970). For comparison with $\rho_{\rm ICR}$ we use both $h = 1$ and $h = 0.75$ and find a density contrast $6 \lesssim \rho_{\rm ICR}/\rho_{\rm LSC} \lesssim 15$.

Now that we have calculated the density of the ICR, we ask if such a large density could result from random fluctuations of a homogeneous "field." The method used to test this hypothesis is given by Chincarini and Rood (1976); we compare the observed redshift distribution of galaxies to that expected from a smooth field. A χ^2 test of the observed redshift distribution of foreground galaxies plus those in the ICR when compared to the theoretical distribution of field

objects (scaled to our larger data sample) shows that there is less than a 10^{-7} probability that the observed distribution arises by chance fluctuations. Although the quoted probability refers to the entire redshift distribution from the foreground out to the Supercluster, it is clear that the ICR dominates the redshift distribution and could not be a chance fluctuation.

c) Properties of the Foreground Galaxies

For the purposes of our analysis we will assume that the foreground consists of those galaxies with $V_0 < 5000$ km s^{-1}. An exception is made for galaxies near the Coma and A1367 centers; for those galaxies with $r < 1°$, a foreground galaxy must have $V_0 < 4500$ km s^{-1}. This variation in the redshift criterion is necessitated by large kinetic energies of the galaxies projected near the centers of massive, relaxed clusters.

Given these criteria, we find eight distinguishable groups in the foreground, and only four out of 90 galaxies do not lie in these eight groups. For seven of these groups, detailed maps and lists of member galaxies are given in the Appendix. In addition to previously published data and new redshifts from Table 1, the lists in the Appendix give new estimates of the morphological types of the component galaxies. These new morphologies were determined by examination of the Kitt Peak National Observatory glass copies of the Sky Survey. In the present section we summarize the general properties of the foreground systems.

In order to compare the intrinsic characteristics of one cluster with those of another, we have extrapolated the mass and luminosity estimates to a fixed absolute limit. We choose $M_p = -15.0$ as the limit because this is just fainter than the faintest galaxy luminosity in the closest group in our survey. Our extrapolation uses the luminosity function introduced by Abell (1975). Ideally, the luminosity function should be derived for each group independently. However, the small number of objects in the individual groups precludes such analysis. Fortunately, the shape of the luminosity function for galaxies in nearby groups seems to be similar to that in rich clusters (see Shapiro 1971).

We define $n(M_{\rm llm})/n(-15)$ to be the fractional number of cluster members that are bright enough to be included in our survey; $M_{\rm llm}$ is the limiting absolute luminosity corresponding to the $m_p = 14.9$ at the distance of each cluster. The total estimated population of the cluster, $N_{\rm calc}$, is then obtained from the relation

$$N_{\rm calc} = N_{\rm obs}[n(M_{\rm llm})/n(-15)]^{-1},$$

where $N_{\rm obs}$ is the number of cluster members with $m_p < 15.0$.

A similar method is used to estimate each cluster's total luminous mass. We define $l_t(M_{\rm llm})/l(-15)$ to be the fractional luminosity of galaxies with $M_p < M_{\rm llm}$. The mass estimate is obtained from

$$\mathfrak{M}_{\rm calc} = \sum_{t={\rm E,S0,S}} N_t(M/L)_t[l_t(M_{\rm llm})/l(-15)]^{-1},$$

TABLE 2

FOREGROUND SYSTEMS

Name	Type	$N_{obs}^{(1)}$	$N_{calc}^{(2)}$	$\bar{V}_0$	σ_V	Morphology (% Spiral)	$\log L_{obs}^{(1,2,3)}$	$\log M_{calc}^{(2,3,6)}$ (4)	(5)	Virial M/L	R_h (degrees)	$\log \rho^{(2,3,4,6)}$ ($M_\odot$ Mpc^{-3})
Coma I	cloud	22	46	933	224	59	11.15	13.25	12.32	(10^3)	2.72	11.63
NGC 4793	cloud	8	35	2513	186	100	10.73	12.77	11.62	1253*	2.74	10.95
NGC 3801	group	14	90	3262	–	65	10.91	13.10	12.12	–	0.58	12.96
NGC 3798	group	3	22	3512	–	(50)	10.42	12.64	11.69	–	0.79	12.00
NGC 3995	group	6	41	3390	131	67	10.85	13.05	12.10	185	0.65	12.71
NGC 4169	group	10	81	3955	140	61	11.10	13.33	12.38	31	0.64	12.81
NGC 4005	group	10	95	4420	–	55	10.95	13.11	12.02	–	0.32	13.35
NGC 4615	cloud	8	73	4623	193	88	11.02	13.17	12.18	247	2.52	10.66

NOTES: (1) $m_p < 15.0$.
(2) calculated for $M_p \le -15.0$
(3) solar units
(4) M/L = 200 for E and S0 galaxies; M/L = 100 for spirals
(5) M/L = 50 for E galaxies, 30 for S0s, and 7 for spirals
(6) assuming H = 75 km s^{-1} Mpc^{-1}

*This value is probably high because of the inclusion of nonmembers; see discussion in text.

where the summation is taken over morphological types grouped into E, S0, and S classes, N_i is the observed number of member galaxies in each class, and $(M/L)_i$ is an assumed mass-to-luminosity ratio for each morphological type. For each foreground system N_{calc} and $\mathfrak{M}_{calc}$ are given in Table 2, and a discussion of the results will be given in § IV. Further discussion of how these results relate to the spectrum of galaxy clustering will be given in a subsequent paper (Gregory and Thompson 1978).

Much of the discussion in § IV will refer to the volume mass densities of the observed clusters. The volume occupied by each cluster is taken to be $V = 4/3\pi R_h^3$, where R_h is the projected mean harmonic radius of the cluster and is found from

$$1/R_h = 1/n_{pairs} \sum_{pairs} 1/r$$

(r is the separation between the members of a galaxy pair; for those pairs with $r < 0°.1$, $1/r$ is not included in the average because large values of $1/r$ force R_h toward unrealistically small values).

For four foreground groups enough data exist to warrant calculation of the virial M/L ratio required to bind the system. For the virial calculation, we use the method described by Materne (1974), except that we assume there is a 100 km s^{-1} uncertainty for the redshift of each galaxy and that redshift differences among group members represent only line-of-sight motion.

In Table 2 we present the observed and calculated data for the foreground systems. The adopted system name and type (group or cloud) are found in columns (1) and (2), respectively. In columns (3) and (4) we list N_{obs} and N_{calc}, while columns (5) and (6) give the mean redshift and dispersion. Column (7) presents the relative number of galaxies with morphological type later than S0 (i.e., %S).[2] The total observed luminosity

[2] Hereafter, %S indicates percent of spirals.

is given in column (8). Columns (9) and (10) contain the mass estimates for two cases. M/L ratios of 200, 200, and 100 were used for E, S0, and S galaxies, respectively, to obtain the values in column (9); M/L ratios of 50, 30, and 7 were used for column (10). Virial M/L ratios are presented in column (11) for the four systems with sufficient data. R_h and the logarithm of the volume mass density ($M_\odot$ Mpc^{-3}) are listed in columns (12) and (13).

Galaxies were assigned membership in groups and clouds on the basis of their projected spatial proximity and their redshift similarities. If clusters merged slowly into a smooth field of galaxies, our subjective method would be unreliable. However, the absence of an observed field makes cluster definition simple and natural when enough redshifts are available. We find that the scale length contrast—i.e., the ratio of the average nearest-neighbor separation of groups and clouds, I_{cl}, to the average mean harmonic radius—is $I_{cl}/R_h = 7$.

In general, the calculated virial M/L ratios are typical for galaxy systems, ranging between $3 \le M/L \le 247$. An exception, however, is the value of $M/L = 1253$ for the NGC 4793 cloud. Such a large value suggests that we may have included nonmembers. However, even after excluding the three most likely nonmembers (NGC 4275, NGC 5089, and IC 777), we find that the ratio is still large, $M/L = 381$. Since the only galaxies in the foreground that might be considered isolated are in the vicinity of this cloud, and since it requires an unusually high mass-to-light ratio to be bound, it seems likely that the NGC 4793 cloud is dispersing.

Considering the general properties given in Table 2 and the details and maps given in the Appendix, we can summarize the foreground region as follows: No evidence is found for a significant population of isolated galaxies. Nearly all galaxies lie in low-mass clusters of the types called "groups" or "clouds" (distinguished by their densities). These groups and clouds have radii which are small compared with their separations. They have low masses, $\mathfrak{M}_{calc} < 2 \times 10^{13}$ $M_{\odot}$, and are sparsely populated, $N_{calc} < 100$. These systems have small velocity dispersions of $\sigma < 200$ km s^{-1}. Of all their component galaxies, 79% are of late morphological type. Finally, these groups have little tendency toward central concentration.

d) Background Galaxies

Because our redshift survey is limited by a fixed apparent magnitude which corresponds to a rather bright intrinsic luminosity $(M_p = -19.9)$ at the distance of the Supercluster, the sample of galaxies lying behind the Supercluster is quite small. Consequently, we can draw only limited conclusions about their distribution. The only background galaxies that were recognized in earlier surveys were too faint to be included in our sample. These galaxies were discussed by Tifft and Gregory (1976), who suggest that they are probably members of groups similar to those in the foreground.

We find evidence for an additional population of background galaxies with $m_p < 15.0$ which lie within 20 Mpc of the Supercluster, having $7900 < V_0 < 9500$ km s^{-1}. These galaxies are easily identified in Figure 3, which is a redshift histogram of those ICR galaxies that do not lie in groups. There are 50 galaxies plotted in the figure with $6000 < V_0 < 9500$ km s^{-1}. An obvious peak is found at the redshift of the Supercluster (6900 km s^{-1}), and a sharp cutoff is found for $V_0 < 6400$ km s^{-1}. We include the NGC 4615 group in Figure 3 to illustrate the large gap of ~ 25 Mpc between the ICR and the highest-redshift group lying in front of the ICR. The redshift distribution of the ICR galaxies is skewed toward higher redshifts. The moment of skewness for all galaxies with $V_0 > 6000$ km s^{-1} is $\gamma = 1.33$ [$\gamma = m_3/(m_2)^{3/2}$, where m_2 and m_3 are the second and third moments about the mean, respectively]. Since no significant skewness is seen in the Coma cluster itself (Gregory and Tifft 1976b), we suggest that the high-redshift tail is caused by the

TABLE 3
BACKGROUND GALAXIES

Zwicky No.	Name	Type	m_p	V_0	Source†
127076S*	N 3926	E	14.7	7464	GT
127076NW	N 3926	E	14.7	8514	GT
158024	N 4104	S0	13.7	8473	CR
158072	N 4272	S0⁻	14.2	8460	CR
158076	IC 3165	Sc	14.9	8428	CR
158100	N 4375	Sab	13.9	9034	CR
159011	N 4514	Sb	14.2	8011	CR
159022W	N 4556	E	14.4	7980	CR
159052		Sc	14.9	9377E	TG
159082		Sb	14.8	8174E	TG

*included because of association with 127076NW

†CR = Chincarini and Rood (1976 + references therein)
GT = present study
TG = Tifft and Gregory (1976 and references therein)

inclusion of ~ 10 background galaxies. However, we cannot rule out the possibility that the ICR has an asymmetric distribution. The curve superposed on the histogram in Figure 3 is a normal distribution fitted to the remaining 40 ICR galaxies.

Before listing the possible background galaxies, we note two instances of CGCG apparent doubles which have one member with $V_0 < 7900$ km s^{-1} (the apparent division between the Supercluster and the background) and one with $V_0 > 7900$ km s^{-1}. Zw 159022w has $V_0 = 7980$ km s^{-1}, and Zw 159022e has $V_0 = 7395$ km s^{-1}. Since these two galaxies show no evidence of physical interaction, we consider them to be a chance optical pair. Zw 127076s has $V_0 = 7464$ km s^{-1} and Zw 127076n has $V_0 = 8514$ km s^{-1}. These two galaxies do seem to share a common envelope, and we place them both in the suggested background, since their mean is $\bar{V}_0 = 7989$ km s^{-1}. In Table 3 we list the 10 galaxies that we consider to be located behind the Supercluster. In columns (1) and (2) the CGCG and NGC/IC identification numbers are given. In column (3) we give the new morphological classifications. Columns (4), (5), and (6) present m_p, the redshift, and the redshift source, respectively.

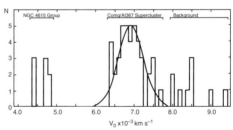

FIG. 3.—The redshift distribution of the isolated ICR galaxies plus the NGC 4615 group and the probable background. There is a very large gap in redshift between the ICR and the highest-redshift galaxies located directly in front of the ICR. The curve is a normal distribution which was fitted to the redshift distribution of the isolated ICR galaxies.

GREGORY AND THOMPSON

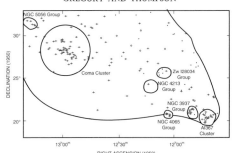

FIG. 4.—A map of the Supercluster. *Dashed curve*, approximate borders of our redshift survey. *Solid curves*, groups and rich clusters. *Crosses*, Supercluster galaxies. *Open circles*, the location of the probable background objects; *small, filled circles*, remaining galaxies with $m_p < 15.0$ which do not have known redshifts.

e) Properties of the ICR Galaxies and the Structure of the Supercluster

In the preceding sections we listed both the foreground and background galaxies that are seen in our sample. The remaining galaxies are part of the Supercluster, and we devote the present section to an examination of their properties and spatial distribution. We identify two previously unrecognized populations of galaxies in the Supercluster that differ from each other and from the rich cluster population in distribution, dynamics, and component morphological types. These two newly recognized populations are (1) galaxies found in intermediate- or low-mass clusters and (2) a dispersed population of isolated galaxies.

In Figure 4 we show the area of the sky that contains the Supercluster. The two rich clusters and the Supercluster groups are outlined by solid curves. The dashed line shows the approximate borders of our survey area. Within the borders there are 31 galaxies with $m_p < 15.0$ that do not have measured redshifts (this number does not include eight unobserved objects which are almost certainly members of the NGC 4005 group). Those galaxies known to be in the foreground have been omitted from Figure 4, so the map contains only the Supercluster galaxies (*crosses*), the 10 probable background galaxies (*open circles*), and the remaining unobserved galaxies (*small filled circles*). We estimate that ~20 of the unobserved galaxies may be members of the Supercluster. Hence our redshift survey is approximately 89% complete in the indicated region of the sky.

The Supercluster groups have average projected densities which are quite high, approximately 9 times that of the isolated ICR galaxies. We also note that the Supercluster groups generally have higher central concentrations than those in the foreground, and there are no systems in the Supercluster comparable to the foreground clouds. Table 4 lists the properties of the individual galaxies in the four supercluster groups; we also include the NGC 5056 group, whose membership

in the Supercluster is doubtful. Although redshifts are not available for a total of eight listed galaxies, they are included with the groups because of their projected proximity to the known group members. In Table 5 we list the properties of the 40 isolated ICR galaxies. (The formats of Tables 4 and 5 are the same as that of Table 3.) The isolated galaxies are not strictly homogeneously distributed throughout the surveyed region. They show a general curving extension from Coma toward A1367 and an avoidance of the region south of the Coma cluster. This indicates that we have encountered the southern boundary to the Supercluster. The sharp boundary on the near side of the ICR is the only other well-determined border.

The isolated ICR galaxies may also have a tendency to congregate near the two rich clusters. To test this suggestion, we grouped these 40 galaxies into the portions of eight concentric annuli that intersect our surveyed region. The annuli were centered on the Coma core and contain five galaxies each. Over the distance range $3° < r < 12°5$ we find that the apparent two-dimensional number density, S, has a shallow falloff away from Coma; $S \propto r^{-0.9 \pm 0.4}$. This result agrees well with that found by Chincarini and Rood (1976), $S \propto r^{-1.2 \pm 0.2}$. A similar calculation based on annuli centered on A1367 yields $S \propto r^{-0.5 \pm 0.1}$ over the distance range $1° < r < 11°5$. We emphasize that these surface density falloffs do not characterize cluster profiles but refer only to the density of galaxies in the ICR (see further discussion in Thompson and Gregory 1978).

We can use these same isolated galaxies to determine whether there is any significant systematic variation of redshift across the Supercluster. Although A1367 has a significantly lower redshift than Coma, the isolated galaxies show only a very marginal trend in the same direction. A linear regression solution of redshift on projected distance along the line of separation of Coma and A1367 gives a result of 10.2 ± 9.7 km s^{-1} degree^{-1} variation from the mean of $\bar{V}_0 = 6902$ km s^{-1}.

TABLE 4

GALAXIES IN SUPERCLUSTER GROUPS

Zwicky No.	Name	Type	m_p	V_0	Source[a]	Zwicky No.	Name	Type	m_p	V_0	Source[a]
(a) NGC 3937 Group						**(c) Zw 128034 Group**					
127063	N 3910	S0	14.4	7768	GT	128004		Sm	14.9	–	
127080	N 3929	E/S0	14.5	7041	GT	128027	N 4101	S0	14.7	6089	GT
127082		Sd	14.7	6555E	DM	128034		S0	14.4	(6700)*	GT
127088	N 3937	S0⁻	14.0	6554	GT	128037	I 762	S0p	14.8	–	
127089	N 3940	S0⁻	14.3	6341	GT	128045		S	14.6	–	
127090	N 3943	SB0	14.7	6559	GT	128047		S0	14.5	–	
127095	N 3947	Sc	14.2	6288E	DM						
127098	N 3954	E	14.4	6800	GT	*bad comparison spectrum					
127100		S	14.9	6840	GT						
127072		Sbc	14.6	–		**(d) NGC 4213 Group**					
097161*	N 3919	E/S0	14.5	6052	GT	128065	N 4213	E⁺	14.3	6986	GT
*membership uncertain						**(e) NGC 5056 Group**					
(b) NGC 4065 Group						160173	N 5056	Sb	13.6	5481	TG
128005	N 4061	E	14.4	7072	GT	160176	N 5057	S0	14.6	5856	TG
128007	N 4065	S0⁻	14.0	6216	GT	160181	N 5065	Sbc	14.3	5732	TG
128008	N 4066	S0	14.4	7294	GT	160183	N 5074	Pec	14.7	5720	TG
128009	N 4070	S0⁻	14.3	7143	GT	160202		–	14.9	5183	TG
128012	N 4076	S0/a	14.3								
128020	N 4089	E:	14.9	6905	GT	161030		S0	14.8	5088	TG
128023	N 4092	Sbc	14.4	6737	GT						
128025	N 4095	S0	14.6	7057	GT						
128026	N 4098	Ep⁺	14.5	7280	GT						
098032	N 4053	S0⁺	14.6	–							
098034		Pec	14.8	–							

[a] DM = Dickens and Moss (1976)
GT = present study
TG = Tifft and Gregory (1976 and references therein)

To find the line-of-sight depth of the Supercluster, we will assume that the redshift dispersion is caused by differential Hubble expansion. This estimated thickness will be only an upper limit because we can *a priori* predict that there will be two other components contributing to the dispersion. One component, σ_{vir}, would be large if these galaxies were ever in virially relaxed groups. Another component, σ_p, represents the primordial random kinetic energy of the galaxies themselves. We assume $\sigma_{vir} \ll \sigma_d$ and $\sigma_p \ll \sigma_d$, where σ_d is the dispersion caused by differential Hubble flow. For an operational definition of the characteristic depth, we use the difference in distances determined by a 2σ spread from the mean. The observed $\sigma = 318$ km s^{-1} then yields a thickness of $6.4\,h^{-1}$ Mpc. Since this is only 30% of the apparent separation between Coma and A1367, the Supercluster is seen to be highly asymmetric. However, if the 10 background galaxies indicate an extension on the far side of the Supercluster, then its depth would be $\sim 13\,h^{-1}$ Mpc.

Table 6 summarizes the general properties of the Supercluster systems; the format is the same as that of Table 2. Four of the eight systems (Coma, two groups, and the isolated ICR galaxies) have mean redshifts in the narrow range $6900 < \bar{V}_0 < 7000$ km s^{-1}. A1367 and the Zw 128034 group have considerably lower values of $\bar{V}_0 = 6450$ and 6395 km s^{-1}, respectively, and the NGC 3937 group has redshifts intermediate between the two extremes. The NGC 5056 group has $\bar{V}_0 = 5510$ km s^{-1}, which is ~ 900 km s^{-1} lower than

any system which is definitely part of the Supercluster. The line-of-sight velocity dispersions vary from $\sigma \approx 300$ to $\sigma \approx 1,000$ km s^{-1}.

The morphological data show significant differences among the populations. Of the foreground galaxies with $M_p \leq M_1$, 69% are spiral or related late types, while only 39% of all the Supercluster galaxies are spiral. Yet, within the Supercluster, we find that 59% of the isolated ICR galaxies are spiral, while spirals account for only 33% and 36% of the galaxies in rich clusters and groups, respectively. Since we are comparing morphologies for galaxies having a wide range in distances and hence diameters, we have compared morphologies judged from the Sky Survey (the source of most of our data) with independent morphologies determined from excellent 4 m prime focus plates of Coma and A1367. We find that the rough statistic of spiral percentage is relatively independent of the plate material and hence also of distance.

For those Supercluster groups where enough data exist, we have calculated the virial M/L ratios; they are found to range $200 < M/L < 400$. N_{calc} and $\mathfrak{M}_{calc}$ were obtained as in § IIIc and show that the Coma cluster, having an estimated 1269 galaxies with $M_p < -15.0$, contributes 43% of the total number of Supercluster galaxies and 37% of the total luminosity. We estimate the total mass of the Supercluster to be $8.4 \times 10^{14}\,M_\odot$ (assuming M/L ratios of 200, 200, and 100 for E, S0, and S galaxies); fully 50% of this mass is located in the Coma cluster itself.

TABLE 5

ISOLATED SUPERCLUSTER GALAXIES

Zwicky No.	Name	Type	m_p	V_o	Source*
127025	N 3808	Sd	14.1	7050	DM
127025B	N 3808	S/S0	--	7230	DM
127038		Sc	14.0	6906	DM
127050		Sa	14.8	6736	GT
127086		E	14.8	6793	GT
127099	N 3951	S	14.5	6430	GT
127115	N 4003	S/S0	14.8	6438	GT
128003		Sd_p	14.6	6476	GT
128017	N 4084	E	14.9	6621	GT
128054		S0⁻	14.6	7220	GT
128077	IC 780	E/S0	14.5	6779	GT
128078	IC 3171	E	14.8	6935	GT
128089	IC 791	Sab	14.2	6735	GT
129011	IC 3582	?	14.3	7113	CR
129012	IC 3581	Scd	14.9	6972	GT
129020	IC 3692	Sc	14.8	6412	GT
129022	IC 813	S_p	14.4	7049E	TG
157054	N 3971	E	13.9	6862	CR
157061	N 3988	E	14.7	6535	CR
158009		S	14.0	7462	CR
156033		E	14.4	6753	CR
156036	N 4146	Sbc	13.8	6461	CR
156053	N 4211	$S0_p$	14.4	6752	CR
158054		S_p	14.6	7689	CR
158081		Pec	14.5	6697	CR
156112	IC 3376	Sd	14.4	7078	CR
159005	IC 3407	Sc	14.7	7062	CR
159021	N 4555	E⁺	13.5	6685	CR
159022E	N 4556	E⁺	14.4	7395	CB
159037	N 4585	S_p	14.6	7411E	TG
159038		$S0_p$	14.6	6972	CR
159059		S_p	14.5	7456E	TG
159061		Sed	14.8	7113E	TG
159070	N 4673	E	13.7	6990	BGC
159072N	N 4676	S_p	14.1	6585	TG, BGC
159072S	N 4676	S_p	--	6598	TG, BGC
159076	IC 821	Sed	14.5	6726	TG, K
159095	IC 826	Sbc	14.9	7153	TG
159103		S0	14.8	6894E	TG
160080		Sb	14.7	6844E	TG

*BGC = de Vaucouleurs and de Vaucouleurs (1964)
 CR = Chincarini and Rood (1976 + references therein)
 DM = Dickens and Moss (1976)
 GT = present study
 K = Kintner (1971)
 TG = Tifft and Gregory (1976 + references therein)

IV. DISCUSSION

In the preceding portions of this paper, we have presented observations dealing with four distinct populations of galaxies that differ from one another in distribution, dynamics, and component morphological types. In the following discussion we will examine the implications which these observations have on the evolution of galaxy systems in general and on the Coma/A1367 Supercluster in particular.

Two of the four galaxy populations have been studied previously.

1. The properties of galaxies in the two rich clusters are known in detail. In addition to the references given in § II, analyses of the properties of these galaxies can be found in Rood et al. (1972), Gregory (1975), and Gregory and Tifft (1976a, b). Here our major objective is to examine the environment in which these clusters are located. We note that the present-day separation between Coma and A1367 is $24\,h^{-1}$ Mpc. Hence

their relative separation (i.e., Hubble flow) velocity is 2400 km s⁻¹ (joint gravitational attraction causes negligible deceleration). The dominant component of this velocity must be in the plane of the sky, since the line-of-sight velocity difference is only 500 km s⁻¹. This indicates that the separation vector between Coma and A1367 lies at an angle of ~12° from the plane of the sky. We find no reason at present to speculate on possible orbital motion of the two clusters about a mutual center of mass.

2. Galaxies in sparse groups, such as those found in the foreground of the Supercluster, have also been studied before, but our findings show some new and unexpected results. Groups are found to have separations much larger than their radii, and the intergroup space is nearly devoid of galaxies. (Chincarini and Martins 1975 have suggested that the redshift distribution in the direction of the Hercules supercluster is also clumpy.) There are large regions of space with radii $>20\,h^{-1}$ Mpc which contain no detectable galaxies, groups, or clusters, giving an upper limit to the detected mass density in these regions of $\rho < 4 \times 10^{-34}$ g cm⁻³. A redshift survey now being done by Gregory, Thompson, and Tifft (1978) which examines the supercluster surrounding A426 (Perseus), A347, and A262 shows that there exist even larger voids than any found in the present study.

It is an important challenge for any cosmological model to explain the origin of these vast, apparently empty regions of space. There are two possibilities: (1) the regions are truly empty, or (2) the mass in these regions is in some form other than bright galaxies. In the first case, severe constraints will be placed on theories of galaxy formation because it requires a careful (and perhaps impossible) choice of both Ω (present mass density/closure density) and the spectrum of initial irregularities in order to grow such large density irregularities. If the second case is correct, then matter might be present in the form of faint galaxies, and an explanation would have to be sought for the peculiar nature of the luminosity function. Alternatively, the material might still be in its primordial gaseous form (either hot or cold neutral hydrogen), and the physical state of this matter may be similar to that discussed in a number of speculative papers (see Rees and Ostriker 1977). A search for radio radiation should be made in the direction of the voids.

a) The Supercluster Groups

Our census of groups within the ICR is probably complete for (1) groups with at least one galaxy having $m_p < 15.0$, and (2) groups with easily recognizable central concentrations. As an example, because the charts in CGCG extend 0.8 mag fainter than our survey, we were able to discern the presence of the NGC 4213 group even though NGC 4213 itself is the only bright galaxy. It will be important to eventually push the redshift survey in the Supercluster to $m_p = 15.7$, the limit of the Zwicky catalog. Undetected groups that do not meet the criteria given above may exist within the Supercluster.

No. 3, 1978

COMA/A1367 SUPERCLUSTER

795

TABLE 6

SUPERCLUSTER SYSTEMS

Name	Type	$N^{(1)}_{obs}$	$N^{(2)}_{calc}$	$\bar{V}_0$	σ_v	Morphology (% Spiral)	$\log L^{(1,2,3)}_{obs}$	$\log \mathcal{M}^{(2,3,4)}_{calc}$ (4)	(5)	Virial M/L	R_h (degrees)	$\log \rho^{(2,3,4,5)}$ ($M_\odot$ Mpc^{-3})
Coma*	rich cluster	66	1269	6947	~1000	27	12.17	14.63	13.87	2×10^3	0.96	12.85
Coma†	cluster core	33	635	6870	944	15	11.89	14.39	13.68		0.43	13.67
A1367†	rich cluster	24	400	6373	715	46	11.73	13.93	13.11	2×10^3	0.45	13.25
NGC 3937	group	9	196	6680	306	44	11.35	13.80	12.97	293	0.44	13.08
NGC 4065	group	11	212	6963	354	15	11.39	13.87	13.08	205	0.35	13.40
Zw 128034	group	6	100	6395	-	33	10.99	13.41	12.55	-	0.62	12.30
NGC 4213	group	1	22	6986	-	0	10.41	12.99	12.39	-	-	-
NGC 5056	group	6	71	5510	316	67	10.94	13.14	12.09	384	0.43	12.70
Isolated ICR	-	40	785	6902	318	59	11.97	14.12	13.29	-	-	-

NOTES: (1) $m_p < 15.0$
(2) calculated for $M_p \leq -15.0$
(3) solar units
(4) M/L = 200 for E and S0 galaxies; M/L = 100 for spirals
(5) M/L = 50 for E galaxies, 30 for S0s, and 7 for spirals
(6) assuming H = 75 km s^{-1} Mpc^{-1}
*out to radius = 3°
†out to radius = 1°

Our estimates of the masses and richnesses of clusters are limited by the accuracy of the assumed luminosity function. For faint galaxies in groups there is some evidence that the slope of the integrated luminosity function might be flatter than the Abell luminosity function used in § IIIc (Gregory and Thompson 1977; Felten 1977). Therefore we might have systematically overestimated N_{calc} and $\mathcal{M}_{calc}$ for Supercluster groups as compared with the foreground groups, since corrections to observed quantities are larger for the more distant systems.

Two independent lines of argument confirm that, even if we have overestimated N_{calc} and $\mathcal{M}_{calc}$ for the Supercluster systems, the Supercluster groups would still be found more massive than the foreground groups. One argument is that the NGC 4065 and NGC 3937 groups each contains at least eight galaxies with $M_p \leq -19.9$, while no foreground system has more than four galaxies that satisfy the same inequality (see Fig. 2). Any reasonable luminosity function would indicate that groups containing many bright galaxies are richer and more massive than systems with few bright galaxies. The second argument is that the line-of-sight velocity dispersions of the NGC 4065 and NGC 3937 groups are higher than any found in the foreground systems. If all of these groups are gravitationally bound, the higher dispersions of the Supercluster groups imply higher masses.

One of the other two Supercluster groups, Zw 128034, also has a mass larger than any of the foreground systems. Unfortunately, five of the brighter galaxies ($m_p < 15.0$) in this system have not had their redshifts determined. It would be useful to see whether its velocity dispersion is also larger than those of the foreground groups. The fourth system, the NGC 4213 group, clearly has a very low mass, but N_{calc} and $\mathcal{M}_{calc}$ are uncertain because only one galaxy is bright enough to be in our survey.

b) The Isolated ICR Galaxies

The only galaxies in our sample that are not located in distinct groups, clouds, or clusters are those we find dispersed within the Supercluster. It is of some importance to determine whether these isolated ICR galaxies are truly primordial "field" galaxies or, alternatively, if they are the remnants of tidally disrupted groups or clouds. If it can be shown that the isolated galaxies are remnants of disrupted groups, then it follows that all galaxies were located within discrete clusters (or groups or clouds) at an early stage of their development. This idea is an alternative to the view that groups, clouds, and clusters grow from small irregularities in the initially smooth galaxy distribution (cf. Press and Schechter 1974; Peebles 1974).

If we hypothesize that the primordial galaxy systems within the ICR were similar to those systems now seen in the foreground sample, the following observations support the idea that dynamic interactions could produce the present-day ICR configuration:

i) The two least-dense foreground systems, the NGC 4615 and NGC 4793 clouds, would be disrupted by Tidal interactions if they were located within $10\,h^{-1}$ Mpc of the Coma cluster or $2.5\,h^{-1}$ Mpc of A1367.

ii) Spirals dominate the least-dense foreground clouds, and we find that the isolated ICR galaxies have the highest spiral incidence among the Supercluster systems.

iii) The isolated galaxies have a weak tendency to congregate near the two rich clusters (see § IIIe). This effect could have two origins. One is that the developing clouds nearest to Coma and A1367 would be the most effectively disrupted. The other is that the two rich clusters attract more than half the galaxies dispersing from a cloud even if the velocity vectors of the galaxies were initially isotropic. (A galaxy located

10 Mpc from the Coma cluster could be completely decelerated from an initial 100 km s^{-1} separation velocity in $\sim 5 \times 10^9$ yr.) This process is reminiscent of the cosmological infall discussed by Gunn and Gott (1972), except the infalling material comes not from a homogeneously distributed "field" but from the highly asymmetric distribution of Supercluster clouds and groups. It is likely that intergalactic gas, if present, would possess a distribution similar to that of the visible galaxies and would therefore also be infalling asymmetrically.

iv) The existing ICR groups are found to lie closer to A1367 than to Coma; this is expected because the more massive cluster should be more effective at disrupting small systems.

Although the present data are not conclusive, it seems likely that the evolution of the structure of the ICR was dominated by tidal interactions with Coma and A1367. We look forward to dynamical simulations of this picture.

c) Epoch of Cluster Formation

If we adopt the conventional viewpoint that groups and clusters were formed by dissipationless collapse, we can use the observations in the present survey to calculate the epoch of cluster formation. Observationally the problem is simple. We determine V_f, the fraction of the total volume which clusters or groups occupy at the present epoch, and then use the relation $(Z_f + 1) = V_f^{-1/3}$ to find the redshift of formation Z_f which is identified with the epoch when the borders of all groups and clusters were in direct contact with one another. Since our total survey volume was selected to include two rich clusters, we will consider only the foreground sample of groups and clouds out to the redshift limit of 5500 km s^{-1}. The small portion of this volume occupied by the groups themselves is

$$V_g = \tfrac{4}{3}\pi \sum (2R_h)^3 \qquad (1)$$

where the sum is taken over all groups, R_h is the harmonic mean radius of each group ($\pi/2$ times the projected harmonic mean radius listed in Table 2), and the factor of 2 in parentheses corrects for the reduction in radius which occurs when a group becomes virialized. If we define $V_f = V_g/V_t$, then

$$V_t = \tfrac{4}{3}\pi \left(\frac{55}{h}\,\mathrm{Mpc}\right)^3 \times \frac{260}{41{,}253}$$

where 260 degree^2 is the area in the sky over which the survey was made. Using the cluster radii from Table 2 in equation (1), we find

$$V_f = \frac{V_g}{V_t} = \frac{1}{549}.$$

This implies that $1 + Z_f \approx 10$. If the cluster-formation process included dissipation, then a tightly bound cluster which appears to have formed at high redshift could have formed more recently. We conclude that the redshift of formation for the foreground groups is $Z_f \lesssim 9$.

d) The Morphologies of Galaxies

The various galaxy systems that are found in our sample show a wide range in the simple morphology index, %S, the fractional number of component galaxies that are spiral or related late type. Figure 5 shows a plot of this morphological index versus mass density of the system (using $M/L = 200, 200, 100$). Two of the galaxy systems that we have identified are not represented in Figure 5 because of the small number of observed galaxies in each. These are the NGC 3798 group and the NGC 4213 group. The isolated ICR galaxies are also not shown in Figure 5 because of difficulty in defining a meaningful mass density. The remaining systems given in Tables 2 and 6 are represented in the following manner: open circles indicate clouds; filled circles indicate foreground

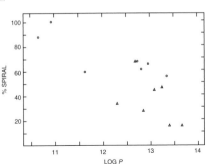

Fig. 5.—A plot of the morphology index versus the logarithm of the volume mass density (in units of $M_\odot$ Mpc^{-3}). The Supercluster systems are represented by solid triangles and are found to have the highest densities and lowest spiral incidence. The foreground clouds (*open circles*) have the lowest densities and highest incidence of spirals. The foreground groups (*small, filled circles*) and the NGC 5056 group (*open triangle*) lie between the extremes of ρ and %S.

groups; filled triangles represent Supercluster systems; the NGC 5056 group is shown as an open triangle.

The data plotted in Figure 5 show a moderate correlation ($r = -0.74$), but the statistical significance of the distribution comes from points with extreme values of ρ and %S. Those systems with intermediate values of ρ show a wide range in %S. It is also significant that the points are segregated according to population. The systems which are called clouds because of their low densities have the highest incidence of spirals. The Supercluster systems have the lowest spiral incidence. It is also interesting that the NGC 5056 group, which lies near the Supercluster but has differing dynamical properties, lies near the foreground groups in this diagram.

e) General Speculations

Our purpose has been to study one supercluster in detail. However, when the Coma/1367 Supercluster is considered along with the other nearby superclusters, at least two features are of note.

Coma and A1367 have a very wide separation, forming a supercluster which is morphologically different from the others. Since important physical processes may manifest themselves by morphological properties, it will eventually be necessary to have a classification scheme for superclusters. A formal scheme should await detailed studies of more exam-

ples, but we can suggest the following three classes on the basis of the nearby examples:

I. *Single core/halo superclusters.*—The prototype is the Local Supercluster, which is centered on Virgo and contains many outlying groups.

II. *Binary superclusters.*—The prototypes are Coma/A1367 (widely separated) and A2197/2199 (nearly in contact).

III. *Extended linear superclusters.*—The prototype for this class is the extensive chain of clusters extending from A426 (Perseus) through A347, A262, and onto NGC 507 and NGC 383 groups.

The final speculation is based on the fact that Coma and A1367 complete an important set of clusters. In Abell's (1958) catalog there are only five clusters with distance class $d \leq 2$ and richness class $\imath \geq 2$. Superclusters containing Hercules, Perseus, and A2199 were previously recognized (e.g., Rood 1976). Coma and A1367 now complete the set. *Every nearby very rich cluster is located in a supercluster.* We suggest that all $\imath \geq 2$ clusters will eventually be found to lie in superclusters. Perhaps such very massive objects can form only in close association with other clusters.

We acknowledge the hospitality of Kitt Peak National Observatory during the observing run and during our subsequent summer visits. S. A. G. received partial support from NSF grant AST 74-22597.

APPENDIX

DEFINITION OF GROUP MEMBERSHIP

The lowest-redshift group is the Coma I cloud. Since Gregory and Thompson (1977) gave a map and a list of member galaxies, we list only its general properties in Table 2. The remaining seven foreground systems are mapped in Figure 6, and we list their member galaxies in Tables 7a–7g; the columns of these tables give the same information as those of Table 3.

The next lowest redshift system after the Coma I cloud has redshifts in the range $2190 < V_0 < 2860$ km s^{-1}. We refer to this system as the NGC 4793 cloud, and it was also recognized by Tifft and Gregory (1976). In Figure 6, the NGC 4793 cloud lies near the eastern boundary of the survey, and Tifft and Gregory list two probable

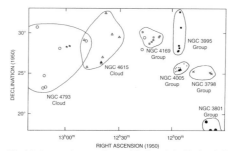

FIG. 6.—A map of seven of the eight foreground systems. (A map of the Coma I cloud is given in Gregory and Thompson 1977.) Two galaxies with redshifts similar to those in the NGC 4793 cloud are seen as open circles projected against the NGC 4169 group. Two other galaxies, lying at the center of the Coma cluster, are shown by star-shaped symbols. These have redshifts similar to those in the NGC 4615 cloud. Small, filled circles within the borders of the NGC 4005 group represent probable members with unknown redshifts.

TABLE 7

GALAXIES IN FOREGROUND GROUPS

Zwicky No.	Name	Type	m_p	V_0	Source*	Zwicky No.	Name	Type	m_p	V_0	Source*
(a) NGC 4793 Group						(e) NGC 4169 Group					
130016	N 5012	Sb	13.6	2566	TG	158029	N 4131	S/S0	14.1	3697	CR
130019	N 5016	Sa	14.3	2769	TG	158030	N 4132	S/S0	14.6	4056	CR
130020		Im	14.8	2592	TG	158031	N 4134	Sbc	13.8	3845	CR
158064	I 777	S	14.5	2607	CR	158041	N 4169	S0	12.9	3829	CR
158073	N 4275	?	13.4	2292	CR	158042		?	14.8	4034	CR
159116	N 4793	Sb	12.3	2504	BGC	158044	N 4174	S/S0	14.3	4150	CR
160134	N 4961	Sb_p	13.5	2582	TG,BGC	158045	N 4175	S	14.2	4050	CR
160194	N 5089	S_p	14.4	2190	TG	158047	N 4185	Sb	13.5	4058	CR
						158050	N 4196	$E/S0_p$	13.7	3966	CR
(b) NGC 3801 Association						158061	N 4253	Sd	13.7	3869	CR
097025	N 3764	Pec	14.9								
097030	N 3768	S/S0	13.7	3288	GT	(f) NGC 4005 Group					
097031	N 3767	SB0	14.5			127101		$S0^-$	14.9		
097032		Sd	14.8			127106	I 746	S	14.5		
097040		S	14.8			127110	N 3987	Sbc	14.4	4508	GT
						127112	N 3993	S0/a	14.8		
097043	N 3790	S/S0	14.5			127114	N 3997	Sd	14.3		
097045		S:	14.6								
097048		S_p	14.3			127120	N 4005	Sa	14.1	4333	GT
097050		SB0	14.6			127122	N 4015	S0	14.2		
097051	N 3801	$S0_p$	13.3	3356	GT	127123	N 4018	S	14.7		
						127125	N 4022	SB0	14.4		
097052	N 3802	S	14.7			127127	N 4023	E/S0	14.6		
097054	N 3806	Sc	14.6								
097067		Sm	14.3								
097070	N 3827	Sc	13.6	3143	DM	(g) NGC 4615 Group					
						129015	N 4614	Sd	14.2	4848	TG,CR
(c) NGC 3798 Group						129018	N 4615	Sd	13.8	4736	TG,CR
127022	N 3798	S	13.9	3459	GT	129025	N 4712	Sc	13.5	4456	CR
127027	N 3812	S0	13.9	3529	GT	159009	N 4495	S	14.1	4361	CR
127064	N 3920	?	14.1	3547	GT	159019		Scd	14.9	4702	CR
(d) NGC 3995 Group						159039		Sc_p	14.0	4375	CR
157060		S0	14.5	3194	CR	159050	I 3651	E	14.4	4777	TG
157065	N 4004	Sc_p	14.0	3420	CR	159092	N 4738	Sbc	14.9	4726	TG,K
157066	N 4008	$S0^-$	13.1	3578	CR						
157068		Sm	14.6	3466	CR						
157069	N 4016	Sd	13.5	3323	CR						
186075	N 3995	Sm	12.9	3356	BGC						

*BGC = de Vaucouleurs and de Vaucouleurs (1964)
 CR = Chincarini and Rood (1976 and references therein)
 DM = Dickens and Moss (1976)
 GT = present study
 K = Kintner (1971)
 TG = Tifft and Gregory (1976 and references therein)

members beyond our borders. Two other galaxies, NGC 4275 and IC 777, are shown in the figure as open circles, the same symbol used for the galaxies whose membership in the cloud is certain. However, these two galaxies lie more than 5° from the rest of the cloud, near the NGC 4169 group. A discussion of the possibility that this system is dispersing can be found in the main body of the paper. Table 7a summarizes the data for the galaxies associated with this cloud.

Between 3000 and 4000 km s^{-1}, the galaxies have a complicated distribution. Chincarini and Rood (1976) grouped all of these galaxies together, but the present evidence shows several different mass concentrations. We first mention the three galaxies shown near the lower right portion of Figure 6. These appear to be part of a loose grouping which we name after NGC 3801. The three members with known redshift have $3143 < V_0 < 3356$. More members of this group may lie to the south of A1367 where no redshifts are available.

About 6° north of the NGC 3801 group are three galaxies with only slightly higher redshifts, $3459 < V_0 < 3547$ km s^{-1}. These are part of a very sparse group that we name after NGC 3798. Five degrees farther north and east we find a line of galaxies extending northward. These six galaxies have redshifts in the range $3194 < V_0 < 3578$ km s^{-1} and will be referred to as the NGC 3995 group, although we note that NGC 3995 is the northernmost member and its physical association with the other five galaxies is not certain.

These last three groups are scattered over 14° (~10 Mpc), but the total range of known redshifts is only 435 km s^{-1}, so they may form a tenuously related association of sparse clusters. The properties of their constituent galaxies are given in Tables 7b, 7c, and 7d.

At somewhat higher redshift we find a much more tightly concentrated cluster which will be referred to as the NGC 4169 group. The redshift range is $3697 < V_0 < 4150$ km s^{-1} with a mean of $\overline{V}_0 = 3955$ km s^{-1} and a dispersion of $\sigma = 133$ km s^{-1}. This is the group that was named after NGC 4131 by Tifft and Gregory (1976), and Chincarini and Rood (1976) referred to it as we do but included galaxies from the NGC 3995 group.

Five degrees southwest of the NGC 4169 group is a concentrated group with only two known redshifts. Since NGC 4005 is the brightest with $m_p = 14.1$, we name the cluster after it. Eight additional galaxies with unknown redshifts are probably associated with this group and are shown as small filled circles in Figure 6.

The highest redshift system that is clearly in the foreground is the NGC 4615 cloud. (The NGC 5056 group is probably not associated with the Supercluster, but we cannot be certain.) Redshifts in this group lie in the range $4361 < V_0 < 4848$ km s^{-1}. Since the lowest redshifts in the center of the Coma cluster itself fall in the upper part of this range, there may be confusion about the membership of individual galaxies. However, the general distribution of low-redshift objects in Coma is very tightly concentrated near the center of the cluster (see Tifft and Gregory 1976, where fainter galaxies show the effect clearly), and the distribution of objects in the NGC 4615 cloud is very loose. Therefore, in general, Coma cluster galaxies can easily be distinguished. Two low-redshift Coma galaxies are shown in Figure 6 with star-shaped symbols. Finally, we point out that the NGC 4615 cloud may be an important object for study, since it is the least-dense cluster, group, or cloud that we have found in this survey.

REFERENCES

Abell, G. O. 1958, *Ap. J. Suppl.*, **3**, 211.
————. 1961, *A.J.*, **66**, 607.
————. 1975, in *Galaxies and the Universe*, ed. A. Sandage, M. Sandage, and J. Kristian (Chicago: University of Chicago Press), p. 601.
Chincarini, G., and Martins, D. 1975, *Ap. J.*, **196**, 335.
Chincarini, G., and Rood, H. J. 1976, *Ap. J.*, **206**, 30.
de Vaucouleurs, G., and de Vaucouleurs, A. 1964, *Reference Catalogue of Bright Galaxies* (Austin: University of Texas Press).
Dickens, R. J., and Moss, C. 1976, *M.N.R.A.S.*, **174**, 47.
Felten, J. 1977, *A.J.*, **82**, 861.
Gregory, S. A. 1975, *Ap. J.*, **199**, 1.
Gregory, S. A., and Thompson, L. A. 1977, *Ap. J.*, **213**, 345.
————. 1978, in preparation.
Gregory, S. A., Thompson, L. A., and Tifft, W. G. 1978, in preparation.
Gregory, S. A., and Tifft, W. G. 1976a, *Ap. J.*, **205**, 716.
————. 1976b, *Ap. J.*, **206**, 934.
Gunn, J. E., and Gott, J. R. 1972, *Ap. J.*, **176**, 1.
Hauser, M. G., and Peebles, P. J. E. 1973, *Ap. J.*, **185**, 757.
Jones, B. J. T. 1976, *M.N.R.A.S.*, **174**, 429.
Kintner, E. C. 1971, *A.J.*, **76**, 409.

Materne, J. 1974, *Astr. Ap.*, **33**, 451.
Noonan, T. W. 1973, *A.J.*, **78**, 26.
Peebles, P. J. E. 1974, *Astr. Ap.*, **32**, 197.
Press, W. H., and Schechter, P. 1974, *Ap. J.*, **187**, 425.
Rees, M. J., and Ostriker, J. P. 1977, *M.N.R.A.S.*, **179**, 541.
Rood, H. J. 1976, *Ap. J.*, **207**, 16.
Rood, H. J., Page, T. L., Kintner, E. C., and King, I. R. 1972, *Ap. J.*, **175**, 627.
Sandage, A. R. 1975, in *Galaxies and the Universe*, ed. A. Sandage, M. Sandage, and J. Kristian (Chicago: University of Chicago Press), p. 761.
Schipper, L., and King, I. R. 1978, *Ap. J.*, **220**, 798.
Shane, C. D. 1975, in *Galaxies and the Universe*, ed. A. Sandage, M. Sandage, and J. Kristian (Chicago: University of Chicago Press), p. 647.
Shapiro, S. L. 1971, *A.J.*, **76**, 291.
Thompson, L. A., and Gregory, S. A. 1978, *Ap. J.*, **220**, 809.
Tifft, W. G., and Gregory, S. A. 1976, *Ap. J.*, **205**, 696.
Tifft, W. G., and Tarenghi, M. 1975, *Ap. J. (Letters)*, **198**, L7.
van den Bergh, S. 1970, *Nature*, **225**, 503.
Zwicky, F., and Herzog, E. 1963, *Catalogue of Galaxies and of Clusters of Galaxies* (Pasadena: California Institute of Technology), Vol. 2 (CGCG).

Note added in proof.—The redshifts reported by Dickens and Moss (1976) are not referred to a galactocentric reference frame as are all other redshifts used in this paper. However, since the galactic rotation corrections are typically only about -60 km s^{-1} for the DM galaxies, our general results are not significantly affected.

STEPHEN A. GREGORY: Department of Physics, Bowling Green State University, Bowling Green, OH   43403

LAIRD A. THOMPSON: Department of Physics and Astronomy, Behlen Laboratory of Physics, University of Nebraska, Lincoln, NE   68508

Notes

1 Understanding the Foundations of Modern Cosmology

1. The cosmological parameters quoted in this book come from the 2015 data release of the European Space Agency's Planck satellite project: Ade et al. (2015). While this paper reports a baryon density of 4.9%, in this book it is rounded up to 5%.
2. Two independent SN Ia teams made this discovery in 1997: Saul Perlmutter (b. 1959) led the first discovery team, and Adam Riess (b. 1969) and Brian Schmidt (b. 1967) shared leadership roles for the second. Other prominent scientists were members of both teams. The necessity of having Lambda as a key component in the cosmological model has now been confirmed by other independent means. Throughout this book, I make the simplifying assumption that the accelerated expansion has its origins with the cosmological constant. Some alternate cosmological models substitute a more general concept called "dark energy" for the cosmological constant.
3. In this book, I do not aim to give a detailed description of Einstein's general relativity, but here I write the basic set of field equations. This might satisfy curious readers who want a taste of the complexities that are involved. Note that subscripts μ and v take on values from 1 to 4 and represent the four components of the space–time manifold. When μ and v sit side-by-side as subscripts, the object formed (written here in bold) is a 4×4 tensor and therefore what appears to be a single equation actually represents a total of $4 \times 4 = 16$ equations. As 6 of the set of 16 equations are redundant, there are only 10 independent relations:

$$\mathbf{R}_{\mu v} - \tfrac{1}{2} a \mathbf{g}_{\mu v} = (8\pi G/c^4)\mathbf{T}_{\mu v} - \mathbf{g}_{\mu v} \text{ Lambda}$$

where
$\mathbf{R}_{\mu v}$ = Ricci curvature tensor
a = numerical value for the curvature of space

$\mathbf{g}_{\mu\nu}$ = metric tensor
G = Newton's gravitational constant
c = speed of light
$\mathbf{T}_{\mu\nu}$ = stress energy tensor
Lambda = cosmological constant

As stated in the text, this equation represents an exact balance between the curvature of space (the left-hand side) and the mass–energy density at that same point (the right-hand side).

4.   In a very practical sense, the Doppler velocity of a galaxy is measured as follows. Galaxy spectra display features that come from those stars that are confined within their boundaries. For example, astronomers often see prominent and easily recognized features of calcium atoms in the spectra of many stars in our Milky Way, and these same features appear in the spectra of galaxies. We know that here on Earth, one particularly well-known calcium feature (Ca K) has a wavelength of 393.366 nm. In the most distant of Slipher's galaxies, NGC 4565, this calcium feature appears not at 393.366 nm but at 394.809 nm, slightly shifted to longer (i.e., redder) wavelengths. The wavelength change is 394.809 nm – 393.366 nm = 1.443 nm. The ratio of this shift to the natural unaltered calcium wavelength is 1.443 nm/393.366 nm = 0.003669, and astronomers call this ratio the "redshift" and label it with the letter "z." If $z \ll 1$, it can be multiplied by the speed of light (c = 299,792 km s^{-1}) to get the so-called Doppler velocity. Slipher obtained a Doppler velocity of 1,100 km s^{-1} for NGC 4565. This works well for the local Universe, but for objects at great distances, the conversion to velocities becomes awkward. Rather than referencing or even discussing the Doppler velocity, astronomers often prefer to use the redshift. This makes sense because, as astronomers probe deeper into the distant Universe, the observed wavelength shifts can *exceed* the natural unaltered wavelength observed here on Earth; z can therefore be greater than 1 and the simple conversion into velocity no longer applies. For example, galaxies have already been detected at $z = 9$. For such large redshifts, the recession velocity approaches the speed of light, and the conversion from the observed z to a velocity must rely on Einstein's theory of special relativity. For those who are curious, the conversion in special relativity is done as follows:

Recession velocity = $c \cdot [(z + 1)^2 - 1]/[(z + 1)^2 + 1]$.
With the example of a galaxy seen at $z = 9$, its recession velocity is $c \cdot 99/101 = c \cdot 0.9802$ or ~98% the speed of light.

5.   Many who hear this for the first time ask, "But how does the Universe expand? Where is the space into which it grows?" This has a simple answer. Instead of saying that the Universe expands, it is equivalent to say that the intrinsic scale (our ruler) used to measure the Universe is shrinking. In this alternate verbalization, space maintains the same volume, but our "rulers" appear to shrink. In his 1933 book *The Expanding Universe*, Eddington was the first to suggest the potential equivalence by discussing "shrinking atoms." For further discussion of subtle complications with this alternate concept, see Yo (2017).

6.   Gamow was an emigrant from Russia who left his homeland shortly before the Communist Revolution. Two great scientists who chose not to immigrate were Andrei Sakharov and Yakov Zeldovich, both of whom would later contribute to the development of Russian nuclear weapons. Sakharov and Zeldovich also wrote scientific papers about the early phases of the Universe, and in their earliest paper, they both assumed the initial state of the Universe was cold. They changed their minds on this issue when the work of Gamow and his students showed great progress, with the alternate hot early Big Bang interpretation.

7.   Often unstated in abbreviated historical accounts of Hubble's work is the fundamental role played by Henrietta Leavitt, who established Cepheid variable stars as "standard candles" in astronomy. This aspect of the story is covered in appropriate detail in Chapter 3 (see Section 3.2).

2   Preview of the Discovery of Cosmic Voids

1.   Early indications of massive halos from flat rotation curves of spiral galaxies came first from observations of Andromeda (M31). These were published by Babcock (1939), Mayall (1951), and Roberts (1966), and the definitive paper with image-intensified spectra by Rubin and Ford (1970). Along the way, Estonian astronomer J. Einasto also contributed to the early analysis. But the general discovery that flat rotation curves dominate in spiral galaxies as a class is most often credited to Rubin, Ford, and Thonard (1978) as well as to Bosma (1978).

2.   Technical definition of the Zeldovich approximation: first-order Lagrangian perturbation theory for the gravitational evolution of initial fluctuations.

3   Homogeneity of the Universe: Great Minds Speak Out

1.   The history of the transcription of Herschel's original observations into Dreyer's New General Catalogue is discussed in detail in a monograph by Steinicke (2010).

2.   van Maanen compared pairs of photographic plates of bright spiral galaxies taken 5 to 15 years apart, using an instrument called a "blink-comparator." He claimed to see measurable rotational shifts of features in the spiral arms on photographic plates taken at different epochs. In reality, bright spiral galaxies rotate on their axes once every ~200 million years, so there is no way for an astronomer to have detected in the early 1900s any rotational shift in the photographic images. His blink comparator had given him a false signal. The complex scientific interplay between the bogus results of van Maanen and the studies of the Universe by other astronomers in that era is commendably described in Smith (1982).

3.   According to the account given by his biographer G. Christiansen, Hubble detected the first extragalactic Cepheid variable star in the Andromeda nebula (M31) in late 1923 and continued to find more Cepheids in observations made

throughout 1924. He notified other astronomers of his discovery by letters he sent out throughout 1924. To be historically proper it must also be noted that a few years earlier, Opik (1922) used a published measurement of the rotation of Andromeda's nucleus and his excellent insight to obtain an accurate distance to Andromeda, before Hubble's discoveries were made.

4. As described by Gingerich (1990), our modern view of the Milky Way as a normal spiral galaxy was clearly elucidated for the first time in 1935 by J. Plaskett when he presented his Oxford University Halley Lecture entitled "The Dimensions and Structure of the Galaxy." Plaskett's work was triggered by a new understanding of interstellar extinction generally credited to Trumpler (1930). The existence of interstellar dust throughout the Milky Way galaxy (dust that produced extinction) confounded Shapley's effort to obtain accurate distances to globular clusters.

5. Just prior to the construction of the 100-inch telescope, Mt. Wilson Observatory's chief optical engineer, George Richey, argued strongly for an advanced optical design (now called the Ritchey-Cretian design) that would have given the 100-inch a much wider field of view. But observatory director George Ellery Hale would not allow it. If this innovative design had been implemented, the history of the large-scale structure might have been significantly different. In the 1930s, the only known faint galaxy clusters were those that happened to be noticed around the edges of long exposure plates on the 100-inch. A wider field of view would have been a big bonus. Reference: D. Osterbrock (1993).

6. The term "metagalaxy" was introduced by Lundmark (1927, Medd. Lund Uppsala Observatory, No. 30) and was used after him by Shapley. In the preface to Shapley's 1957 book entitled *The Inner Metagalaxy*, he explains that this word refers to the measurable material universe, including the assemblage of galaxies as well as the gas, particles, planets, stars, and star clusters in the spaces between the galaxies. Metagalaxy is no longer used in astronomy except in a historical context.

7. One of these rare occasions was in 1934, where Hubble discussed the Shapley and Ames Catalogue, stating that Shapley and Ames recognized "the strong clustering in the northern galactic hemisphere, and the general unevenness of the distribution," and soon thereafter, Hubble says that Shapley "further emphasized the apparent irregularities in distribution and the greater richness in the northern hemisphere."

8. As described in Chapter 4, Fritz Zwicky concurred with Hubble in his views on the galaxy distribution. Zwicky continued to advocate this same homogeneous model as late as 1972, two years before his death in 1974. In the introduction to Hubble's "The Realm of the Nebulae," he states "The views presented here are the shared views of Zwicky and Tolman...." Hubble (1936b).

9. In his book *The Large Scale Structure of the Universe*, Peebles (1980) discussed the history of this era and reports "Shapley's remarks did not attract much attention. ... by the 1950s, the possibility of large-scale inhomogeneity was largely displaced in the minds of cosmologists by the debate over homogeneous world models."

10.   For example, Hubble's colleague Prof. Fritz Zwicky discussed in this era the "tired light theory" in which light loses energy (and, therefore, decreases its frequency) as it passes over long distances through the Universe. This is a theory that has long since been abandoned. Hubble makes no reference to any specific explanation for the redshift phenomenon in his paper.

4   All-Sky Surveys in the Transition Years 1950–1975

1.   The description presented here of the history of our Local Supercluster is very much abbreviated. To see a more complete discussion, another good source is Rubin (1989).

5   The Early Redshift Surveys from Arizona Observatories

1.   The referee for the Coma/A1367 Supercluster paper chose to remain anonymous. However, he/she asked us to add to our manuscript a reference to Chincarini and Martins (1975), the relatively obscure paper on Seyfert's Sextet that had become Chincarini's obsession. This Seyfert's Sextet paper is poorly known and is not often cited. The only astronomer I know from this era who would make this request is Chincarini himself. On the other hand, Rood and Chincarini worked so closely together that their work and their opinions (even when they each refereed papers) were difficult to separate. If Rood was the referee, Chincarini might have influenced him to insist on adding the Chincarini and Martins (1975) reference. This speculation needs confirmation by a historian of science, in the future. Even if Chincarini was not the referee, he attended the IAU Symposium No. 79 where Tifft discussed in mid-September 1977 the Coma/A1367 results and the voids that Gregory and I had discovered months before Chincarini submitted his paper to *Nature*. Chincarini references the Tifft and Gregory conference presentation at Tallinn in his paper to *Nature*, and that is fair enough. However, Chincarini was certainly not the first astronomer to detect cosmic voids as he claims in Chincarini (2013). The Gregory and Thompson paper defined the cosmic void phenomenon, and the first confirmation came from Gregory, Thompson, and Tifft (1981), the Perseus supercluster study that was in a preliminary form in September 1977. It was significant enough at that time to be referenced in the Gregory and Thompson (1978) paper. Chincarini developed his mathematical test for field galaxies on his own over a number of years; that aspect of his 1978 *Nature* paper is legitimate. But his use of the Hercules supercluster data set in his *Nature* paper did not adhere to standard scientific protocol.

2.   Chincarini carries with him a memory of talking with his colleague John Cowan at the University of Oklahoma about the use of the words "hole" and "void" regarding the 3D configuration of the galaxy distribution. Rood (1988b) told the same story. Both say that they talked with Cowan while they were writing their popular article entitled *The Cosmic Tapestry* published in *Sky and Telescope*, May 1980. I presume this conversation took place, but what Chincarini and Rood have overlooked in their discussions is the fact that Gregory and I used the word "void" in our Coma/A1367 Supercluster paper with a manuscript submission date of September 7, 1977. Tifft also used the work "void" in his Tallinn conference presentation when referring to the Coma/A1367 study because we requested him to do so. These events occurred before Chincarini and Rood (1978) began to write "The Cosmic Tapestry" popular article in 1978. I was collaborating with Chincarini and Rood in those days, and I know that they did not begin to write their *Sky & Telescope* paper until spring 1978.

3.   This was told to me by Jaan Einasto, who happened to be visiting Yale University at the time the public relations campaign was playing out. Yale was the home institution of the Boötes void paper's second author, Oemler.

4.   Two separate facts must be added. First, the reference Oort makes in his review article to Giovanelli, Haynes, and Chincarini (1986) is given as *1983 in preparation*. Second, it is notable that Oort excludes from his review four papers discussed in this book: Chincarini and Martins (1975), Chincarini and Rood (1975), Chincarini and Rood (1976), and Chincarini (1978).

6   Galaxy Mapping Attempt at Tartu Observatory

1.   Einasto describes the detector as an "optical multichannel analyzer" but says no more about it. However, based on information in the paper by Luud et al. (1978) it is clear that Tartu Observatory purchased a Silicon Intensified Target (SIT) Vidicon detector system from Princeton Applied Research Corporation that was, indeed, operated as an optical multichannel analyzer.

2.   The discussion in Einasto (2014) is on pp. 139–140. It suggests that the astronomy journals show prejudice against pioneering work in general. While that may be partially true, what he fails to acknowledge is that journal editors are required to scrutinize all manuscripts to identify ideas that are not sufficiently substantiated by solid observations. To be specific, the Tartu Observatory group could not prove that the "holes" they reported in the galaxy distribution were actually empty, nor could they prove that the walls of the cells (that they claimed surrounded the empty cells) were Zeldovich pancakes. When Gregory and I presented our evidence for cosmic voids and for a bridge of galaxies connecting the Coma cluster with A1367, we had little to no trouble getting our manuscript approved by the editor at the *Astrophysical Journal*. Judgments like those given to the Tartu Observatory group are not prejudicial if the authors are making claims based on inadequate evidence. Even at IAU Symposium No. 79, meeting attendee Joseph Silk recognized the potential problem of incompleteness in the samples that were being used by Einasto

and Jõeveer. These are very practical matters and are not necessarily the result of prejudice.

3. While the two IAU Symposium No. 79 papers by Tully (1978a, 1978b) freely use the term "void" when describing the galaxy distribution, it is important to note that all participants wrote their conference reports after the Tallinn meeting was finished, and therefore, after Tifft discussed in a public way the Gregory and Thompson results in Tallinn. It was during this conference that the word "void" entered the jargon of cosmology. Tully's conference contribution discussing the Local Void was neither completed nor submitted to the conference organizers for publication before the Gregory and Thompson Coma/A1367 manuscript arrived at the *Astrophysical Journal* on September 7, 1977.

7   Theoretical Models of Galaxy Formation – East versus West

1. The Moscow school originally suggested that pancake formation would occur at redshifts of $z \sim 4$ to 5, while the most distant sources identified in the early 1970s were quasars at $z = 2.5$. Today, astronomers see individual objects at least to $z \sim 9$, and if these pancakes indeed existed, they could be detected with optical and X-ray telescopes available today (circa 2020) but not when they were first proposed in the early 1970s.

2. The distance scale of the Universe is set by the Hubble constant. In both Lemaître's 1927 analysis and Hubble's 1929 analysis of the galaxy velocity-distance relation, the Hubble constant was claimed to be ~500 km s^{-1} per Megaparsec whereas today, the Hubble constant is known to be (within a few percent) 70 km s^{-1} per Megaparsec. To understand how the scale impacts the hierarchical theory, consider a galaxy that is observed to show a redshift in its spectrum of 1000 km s^{-1}. Hubble, Lemaître, and Holmberg would say it lies at a distance of 2 Mpc [(1,000 km s^{-1}) / (500 km s^{-1} Mpc^{-1})]. Today, however, this galaxy is judged to be at a distance = 1,000 / 70 ~14 Mpc. In other words, the distances to all the galaxies, as deduced today, are greater by the factor 500/70 ~seven times (the ratio of the old and the new values for the Hubble constant) compared to the distances used in the mid-1930s. The Hubble constant is always under close scrutiny, and it has been adjusted many times in the last 90 years as astronomers reassess the cosmic distance scale. Because of these revisions, the Universe appears today to be significantly less dense (fewer galaxies per unit volume) than it did in the 1930s when Holmberg and others first discussed the formation of groups and clusters of galaxies. A decrease in the density of galaxies reduces the chances of random galaxy–galaxy collisions, and this makes the estimated rate of hierarchical growth much lower today when compared to estimates made in the 1930s.

3. ITEP: Institute of Theoretical and Experimental Physics of the USSR Academy of Sciences, Moscow, USSR.

4. Despite the enthusiasm of the Russian group, limits had already been placed on the viability of massive neutrinos to explain the dark matter halos of low-mass galaxies by Tremaine & Gunn (1979).

5. Doroshkevich and Shandarin (1978) do not describe their numerical modeling details but Doroshkevich et al. (1980) reference a book by Hockney published in 1970. The 1980 paper states that they tracked 4,096 particles embedded in 64 × 64 cells in 2D. The Doroshkevich et al. paper appeared in a Russian journal dated September 1980. Melott's new computer model was described in a publication he submitted to the *Astrophysical Journal* on 19 October 1981 as Melott (1983). This paper references a 1981 book by Hockney and Eastwood, *Computer Simulation Using Particles*.

6. IREX, International Research and Exchange Board sponsored scholarly exchange between the US and the Soviet Union before the fall of the Iron Curtain.

7. In the CfA1 paper, Davis, Huchra, Latham, and Tonry (1982) are very frank when describing their "broad but shallow" redshift survey. The CfA1 survey was a useful contributor to delineating the large-scale structure in the galaxy distribution only when combined with the "narrow but deep" redshift surveys from our Arizona surveys. This point was clearly stated in the Introduction of their paper where they compared their shallow redshift survey to our more detailed but narrower redshift surveys.

8. An excellent and succinct description of the link between initial irregularities and inflation is given by Longair (2006, pp. 447–448). This description includes references to Alan Guth's historical discussion of the original contribution by Gibbons and Hawking (1977). Significantly before Gibbons, Hawking, and Guth showed that an inflationary model for the Universe could generate a scale-free distribution of adiabatic initial irregularities, E. Harrison and Y. Zeldovich speculated that such a scale-free spectrum sat at the foundation of galaxy formation. The original papers were Harrison (1970) and Zeldovich (1972).

9. The h in this formula represents the scaled Hubble expansion parameter so that $H = h \times 100$ km s^{-1} Mpc^{-1}. The currently accepted value of H = 70 km s^{-1} Mpc^{-1} implies that $h = 0.7$. Using $h = 0.7$, the DEFW 1985 test volume was clearly stated to be 32.5 h^{-1} ~46 Mpc. In such a volume, one might expect to find one large supercluster and perhaps one to two cosmic voids. This is a volume somewhat smaller than that included in the Gregory and Thompson (1978) study of the Coma/A1367 supercluster. In WFDE 1987, the test volumes are given with fixed dimensions (h is not mentioned) as 280 Mpc and 360Mpc on a side. As best I deduce, by carefully inspecting their paper, they seem to have assumed $h = 0.50$, so I will write the dimension as 140 h^{-1} Mpc and 180 h^{-1} Mpc and then scale to the current $h = 0.7$ to get 200 Mpc and 256 Mpc on a side for their actual test volumes. These are significantly larger than the Gregory and Thompson (1978) Coma/A1367 redshift survey volume.

10. P. J. E. Peebles was interviewed by the *New York Times* science writer Dennis Overbye for a March 1, 2003 article entitled "Universe as a Donut: New Data, New Debate." By the time of this interview, the WMAP probe that studied the CMB had confirmed many aspects of the CDM model: the high-density flat Universe and the slightly "tilted" spectrum of initial irregularities. Overbye

quotes Peebles as having said "Cosmologists have built a house of cards, and it stands."

8   Priority Disputes and the Timeline of Publications

1. Most likely, the referee selected for the de Lapparent, Geller, and Huchra (1986) paper was from one of two groups: those astronomers who were associated with Marc Davis at UC Berkeley or one of the Princeton University "holdouts for homogeneity." Both of these groups had, like Geller and Huchra, seemingly fallen into the habit of simply ignoring the contributions from the early pioneering redshift surveys from Arizona.
2. In the twenty-four-year period discussed here (1986–2010), four papers were published by other first authors (but with Geller as a secondary author) that do reference the Gregory and Thompson 1978 Coma/A1367 discovery paper. However, these are not part of what might be called the "classic" Geller and Huchra redshift survey papers that often are broadly (and erroneously) given credit for being the first wide-angle galaxy redshift survey.
3. Geller had no basis for suggesting in her interview that current theories (ca. 1991) were unable to explain the structure. For example, Peebles (1982a) was the first of many to suggest how cosmic voids are dynamically emptied, a process that eventually leads to a sharply defined void border where galaxies accumulate. More extended structures in the galaxy distribution arise naturally in galaxy formation models that include the concepts introduced and developed by Zeldovich, Doroshkevich, and Shandarin.
4. While displaced credit was the key issue in both situations, for the redshift survey work, the controversy involved the discovery observations themselves. For the CMB, the controversy involved theoretical predictions by Alpher and Herman that preceded the actual discovery. Long after the events unfolded, Alpher and Herman (2001) described their work in detail.
5. To those who look today with 20/20 hindsight at the 1975 and 1976 Chincarini and Rood Coma cluster redshift plots and say that they can see cosmic voids in the data, I present the following analogy. In 1802, Wollaston noticed dark lines in the solar spectrum and in 1814, Fraunhofer did the same. No one could imagine Wollaston or Fraunhofer claiming that they had discovered the Bohr atom just because their spectra showed an orderly set of lines, the nature of which neither of them understood. This is the nature of precursors to new discoveries and to new paradigms. Like Wollaston and Fraunhofer, Chincarini and Rood did not realize the deeper significance of their own observations until late in 1977, after Gregory and Thompson (1978) had submitted the Coma/A1367 Supercluster manuscript for publication and began to discuss the true physical meaning of the observed gaps in the redshift distribution of galaxies.

9   Impact of Cosmic Voids: Cosmology, Gravity at the Weak Limit, and Galaxy Formation

1. To astronomers, "photometry" means to measure an object's brightness at a specific selected wavelength. Photometric redshifts rely on multiple brightness measurements of the target galaxy at a number of (say, six) different specific wavelengths. These six brightness measurements, as long as they are each sufficiently precise, reveal the target galaxy's overall spectral shape. This shape is then used to determine the approximate redshift of the target galaxy. Wide-field cameras available today can simultaneously image thousands of target galaxies in one pointing. With six such images – each at a different wavelength – photometric redshifts can be simultaneously determined for nearly all target galaxies in each of the selected fields of study.

2. In 2005, a group of astronomers led by D. Eisenstein announced the detection of what is called the "Baryonic Acoustic Oscillation" (BAO) signal. The BAO is a subtle perturbation in the galaxy-to-galaxy separation that arises from sound waves (hence the use of the word "acoustic") that naturally propagate in the early Universe. The BAO test solidly confirms the outward acceleration of the Universe produced by the negative gravitational effect of the cosmological constant and/or the so-called dark energy. See D. Eisenstein et al. (2005).

3. Kirshner, Oemler Jr., Schechter, and Shectman (1981) originally claimed to have discovered a "Million Cubic Megaparsec Void," but their 1987 follow-up paper reports a more accurate void radius of 31.5 h^{-1} Mpc. If we use a modern value for the Hubble constant of 72 km^{-1} Mpc^{-1}, the volume of the Boötes void is ~385,000 Mpc^3, falling short of the number quoted in the 1981 paper. This adjusted volume is large enough that exaggeration is unnecessary: it is an impressive void.

4. Hoffman and Shaham (1982); Fujimoto (1983); Hausman, Olson, and Roth (1983); Hoffman, Salpeter, and Wasserman (1983); Icke 1984; Bertschinger (1985).

5. References to redshift space distortion (RSD) research results follow: Padilla, Ceccarelli, and Lambas (2005); Paz et al. (2013); Micheletti and 48 other authors (2014).

Biographical Sketches

References

Aarseth, S., Gott, J., III, & Turner, E. (1979). N-Body Simulation of Galaxy Clustering. I. Initial Conditions and Galaxy Collapse Times. *Astrophys. J.*, 228, pp. 664–83.

Abbott, T. et al. (2018). Dark Energy Survey Year 1 Results: Cosmological Constraints from Galaxy Clustering and Weak Lensing. *Phys. Rev. D*, 98, Article 043526.

Abell, G. (1958). The Distribution of Rich Clusters of Galaxies. *Astrophys. J. Suppl.*, 3, pp. 211–88.

Abell, G. (1961). Evidence Regarding Second-Order Clustering of Galaxies and Interactions Between Clusters of Galaxies. *Astron. J.*, 65, pp. 607–13.

Abell, G. (1977). Interview of George Abell by Spenser Weart on November 14, 1977, Niels Bohr Library & Archives, American Institute of Physics, College Park, MD USA, www.aip.org/history-programs/niels-bohr-library/oral-histories/4475

Abell, G. & Chincarini, G. (1983). *Early Evolution of the Universe and Its Present Structure* (Dordrecht, Netherlands: D. Reidel Publishing).

Abell, G., Morrison, D., & Wolff, S. (1994). *Realm of the Universe*, Fifth Edition (Ft. Worth, TX: Saunders College Publishing).

Ade, P. A. R., Aghanim, N., Arnaud, M., et al. (2016). Planck 2015 Results. XIII. Cosmological Parameters. *Astron. & Astrophys.*, 594, A13, pp. 1–63.

Alcock, C. & Paczynski, B. (1979). An Evolution Free Test for Non-zero Cosmological Constant. *Nature*, 281, pp. 358–9.

Alpher, R., Bethe, H., & Gamow, G. (1948). The Origin of Chemical Elements. *Phys. Rev.*, 73, pp. 803–4.

Alpher, R. & Herman, R. (1948). Evolution of the Universe. *Nature*, 162, pp. 774–5.

Alpher, R. & Herman, R. (2001). *Genesis of the Big Bang* (New York, NY: Oxford University Press, Inc.).

Aragon-Calvo, M. & Szalay, A. (2013). The Hierarchical Structure and Dynamics of Voids. *Mon. Not. Royal Astron. Soc.*, 428, pp. 3409–24.

Arp, H. (1973). Neighborhoods of Spiral Galaxies. I. Multiple Interacting Galaxies. *Astrophys. J.*, 185, pp. 797–808.

Baade, W. (1951). Galaxies – Present Day Problems. *Publ. of the Observatory of Michigan*, 10, pp. 10–7.

Babcock, H. (1939). The Rotation of the Andromeda Nebula. *Lick Obs. Bull.* (No. 498), 19, pp. 41–51.

Bahcall, J. & Joss, P. (1976). Is the Local Supercluster a Physical Association? *Astrophys. J.*, 203, pp. 23–32.

Baldry, I. (2008). What Hubble Really Meant by Late and Early Type: Simply More or Less Complex in Appearance. *Astron. & Geophys.*, 49, pp. 25–6.

Balzano, V. & Weedman, D. (1982). Filling the Void in Bootes. *Astrophys. J. Lett.*, 255, pp. L1–L4.

Bahcall, N. & Soneira, R. (1982). An Approximately 300 Mpc Void of Rich Clusters of Galaxies. *Astrophys. J.*, 262, pp. 419–23.

Bahcall, N. & Soneira, R. (1984). A Supercluster Catalog. *Astrophys. J.*, 277, pp. 27–37.

Batuski, D. & Burns, J. (1985). Finding Lists of Candidate Superclusters and Voids of Abell Clusters. *Astron. J.*, 90, pp. 1413–24.

Beacom, J., Dominik, K., Melott, A., Perkins, S., & Shandarin, S. (1991). Gravitational Clustering in the Expanding Universe: Controlled High-Resolution Studies in Two Dimensions. *Astrophys. J.*, 372, pp. 351–63.

Berendzen, R., Hart, R., & Seeley, D. (1984). *Man Discovers the Galaxies* (New York, NY: Columbia University Press).

Bernheimer, W. (1932). A Metagalactic Cloud between Perseus and Pegasus. *Nature*, 130, p. 132.

Bertschinger, E. (1985). The Self-Similar Evolution of Holes in an Einstein-De Sitter Universe. *Astrophys. J. Suppl.*, 58, pp. 1–37.

Bertschinger, E. & Dekel, A. (1989). Recovering the Full Velocity and Density Fields from Large-Scale Redshift-Distance Samples. *Astrophys. J. Lett.*, 336, pp. L5–L8.

Beygu, B. Peletier, R., Van der Hulst, J., Jarrett, T., Kreckel, K., Van de Weygaert, R., Van Gorkom, J., & Aragon-Calvo, M. (2017). The Void Galaxy Survey: Photometry, Structure and Identity of Void Galaxies. *Mon. Not. Royal Astron. Soc.*, 464, pp.666–79.

Blumenthal, G., Pagels, H., & Primack, J. (1982). Galaxy Formation by Dissipationless Particles Heavier than Neutrinos. *Nature*, 299, pp. 37–8.

Blumenthal, G., Faber, S., Primack, J., & Rees, M. (1984). Formation of Galaxies and Large-Scale Structure with Cold Dark Matter. *Nature*, 311, pp. 517–25.

Blumenthal, G., Da Costa, L., Goldwirth, D., Lecar, M., & Piran, T. (1992). The Largest Possible Voids. *Astrophys. J.*, 388, pp. 234–41.

Bok, B. (1934). Apparent Clustering of Galaxies. *Nature*, 133, p. 578.

Bok, B. (1978). Harlow Shapley. *Natl. Acad., Sci.., Biographical Memoirs*, 49, pp. 238–91.

Bond, J., Efstathiou, G., & Silk, J. (1980). Massive Neutrinos and the Large-Scale Structure of the Universe. *Phys. Rev. Lett.*, 45, pp. 1980–4.

Bond, J., Kofman, L., & Pogosyan, D. (1996). How Filaments Are Woven into the Cosmic Web. *Nature*, 380, pp. 603–6.

Bond, J., Szalay, A., & Turner, M. (1982). Formation of Galaxies in a Gravitino-Dominated Universe. *Phys. Rev. Lett.*, 48, pp. 1636–9.

Bondi, H. (1947). Spherically Symmetric Models in General Relativity. *Mon. Not. Royal Astron. Soc.*, 107, pp. 410–25.

Bondi, H. (1948). The Steady-State Theory of the Expanding Universe. *Mon. Not. Royal Astron. Soc.*, 108, pp. 252–70.

Bosma, A. (1978). The Distribution and Kinematics of Neutral Hydrogen in Spiral Galaxies of Various Morphological Types. Ph.D. thesis (Groningen, Netherlands: Groningen University).

Bothun, G., Beers, T., Mould, J., & Huchra, J. (1985). A Redshift Survey of Low-Surface-Brightness Galaxies. I – The Basic Data. *Astron. J.*, 90, pp. 2487–94.

Bothun, G., Beers, T., Mould, J., & Huchra, J. (1986). A Redshift Survey of Low Surface Brightness Galaxies II. Do They Fill the Voids? *Astrophys. J.*, 308, pp. 510–29.

Burbidge, E. (2002). Gerard de Vaucouleurs 1918–1995: A Biographical Memoir. *Natl. Acad., Sci., Biographical Memoirs*, 82, pp. 1–17.

Catapano, P. (2015). Massimo Tarenghi: A Lifetime in the Stars. *CERN Courier*, 55 (August Issue), pp. 31–3.

Centrella, J. & Melott, A. (1983). Three-Dimensional Simulation of Large-Scale Structure in the Universe. *Nature*, 305, pp. 196–8.

Chandrasekar, S. & Munch, G. (1952). The Theory of the Fluctuations in Brightness of the Milky Way. V. *Astrophys. J.*, 115, pp. 103–23.

Charlier, C. (1908). Arkiv fur Matematik, *Astronomi och Fysik*, 4, p. 1.

Charlier, C. (1922). Arkiv fur Matematik, *Astronomi och Fysik*, 16, pp. 1–37.

Chincarini, G. (1978). Clumpy Structure of the Universe and General Field. *Nature*, 272, pp. 515–6.

Chincarini, G. (2013). Large-Scale Structure: The Seventies & Forty Years Later: From Clusters to Clusters. In *The Thirteenth Marcel Grossmann Meeting*, eds. K. Rosquist, R. Jantzen, & R. Ruffini. (Singapore and Teaneck, NJ: World Scientific Publishing Company), also arXiv:1305.2893.

Chincarini, G. & Martins, D. (1975). On the "Seyfert Sextet," VV 115. *Astrophys. J.*, 196, pp. 335–7.

Chincarini, G. & Rood, H. (1975). Size of the Coma Cluster. *Nature*, 257, pp. 294–5.

Chincarini, G. & Rood, H. (1976). The Coma Supercluster – Analysis of Zwicky-Herzog Cluster 16 in Field 158. *Astrophys. J.*, 206, pp. 30–7.

Chincarini, G. & Rood, H. (1980). The Cosmic Tapestry. *Sky & Telescope*, 59, pp. 364–7.

Chincarini, G., Thompson, L., & Rood, H. (1981). Supercluster Bridge between Groups of Galaxy Clusters. *Astrophys. J. Lett.*, 249, pp. L47–L50.

Christianson, G. (1995). *Edwin Hubble: Mariner of the Nebulae* (Chicago, IL: The University of Chicago Press).

Colberg, J., Pearce, F., Foster, C., et al. (2008). The Aspen-Amsterdam Void Finder Comparison Project. *Mon. Not. Royal Astron. Soc.*, 387, pp. 933–44.

Cowsik, R. & McClelland, J. (1973). Gravity of Neutrinos of Nonzero Mass in Astrophysics. *Astrophys. J.*, 180, pp. 7–10.

Cromwell, R. & Weymann, R. (1970). Changes in the Nuclear Spectrum of the Seyfert Galaxy NGC 4151. *Astrophys. J. Lett.*, 159, pp. 147–50.

Croswell, K. (2001). *The Universe at Midnight: Observations Illuminating the Cosmos* (New York, NY: The Free Press).

Davis, M. (1988). Interview of Marc Davis by Alan Lightman on 14 October 1988 Niels Bohr Library & Archives, American Institute of Physics, College Park, MD USA, www.aip.org/history-programs/niels-bohr-library/oral-histories/34298

Davis, M. (2014). Cosmic Structure. *International J. Modern Phys. D*, 23, article number 1430021 (Ch. 14 in Vol. 2, *One-Hundred Years of General Relativity* ed. Wei-You Ni (Singapore: World Scientific Publ. 2014).

Davis, M. Efstathiou, G., Frenk, C., & White, S. (1985). The Evolution of Large-Scale Structure in a Universe Dominated by Cold Dark Matter. *Astrophys. J.*, 292, pp. 371–94. (DEFW)

Davis, M., Huchra, J., Latham, D., & Tonry, J. (1982). A Survey of Galaxy Redshifts. II. The Large Scale Space Distribution. *Astrophys. J.*, 253, pp. 423–45.

Dick, S. (2013). *Discovery and Classification in Astronomy: Controversy and Consensus* (Cambridge, UK: Cambridge University Press).

Dicke, R., Peebles, P., Roll, P., and Wilkinson, D. (1965). Cosmic Black-Body Radiation. *Astrophys. J.*, 142, pp. 414–9.

Doroshkevich, A. & Shandarin, S. (1978). A Statistical Approach to the Theory of Galaxy Formation. *Soviet Astron.*, 22, pp. 653–60.

Doroshkevich, A., Shandarin, S., & Saar, E. (1978). Spatial Structure of Protoclusters and the Formation of Galaxies. *Mon. Not. Royal Astron. Soc.*, 184, pp. 643–60.

Doroshkevich, A., Kotok, E., Poliudov, A., Shandarin, S., Sigov, I., & Novikov, I. (1980). Two Dimensional Simulation of the Gravitational System Dynamics and Formation of the Large-Scale Structure of the Universe. *Mon. Not. Royal Astron. Soc.*, 192, pp. 321–37.

Doroshkevich, A., Zeldovich, Y., & Sunyaev, R. (1976). Adiabatic Theory of Formation of Galaxies. In Origin and Evolution of Galaxies and Stars,OEGS Conference, pp. 65–104. (in Russian).

Doroshkevich, A., Zeldovich,,Y., Sunyaev, R., & Khlopov,M. (1980a). Astrophysical Implications of the Neutrino Rest Mass. II. The Density Perturbation Spectrum and Small-Scale Fluctuation in the Microwave Background. *Astron. Lett. (Russian)*, 6, pp. 457–64.

Doroshkevich, A., Zeldovich,, Y., Sunyaev, R., & Khlopov, M. (1980b). Astrophysical Implications of the Neutrino Rest Mass. III. The Non-Linear Growth of Perturbations and Hidden Mass. *Astron. Lett. (Russian)*, 6, pp. 465–9.

Eddington, A. (1923). *The Mathematical Theory of Relativity*. (Cambridge, UK: Cambridge University Press).

Eddington, A. (1934). Messenger Lectures; (1935) New Pathways in Science.

Efstathiou, G., Davis, M. White, S., & Frenk, C. (1985). Numerical Techniques for Large Cosmological N-body Simulations. *Astrophys. J. Suppl.*, 57, pp. 241–60. (EDWF)

Einasto, J. (2014). *Dark Matter and Cosmic Web Story* (Singapore: World Scientific Publishing).

Einasto, J. (2018). Cosmology Paradigm Changes. *Ann. Rev. Astron. & Astrophys.*, 56, pp. 1–39.

Einasto, J., Joeveer, M., & Saar, E. (1980). The Structure of Superclusters and Supercluster Formation. *Mon. Not. Royal Astron. Soc.*, 193, pp. 353–75.

Einstein, A. (1915). The Field Equations of Gravitation. *Transactions of the Royal Prussian Academy of Sciences (Berlin)* pp. 844–7 (in German).

Einstein, A. (1917). Cosmological Reflections on General Relativity. *Transactions of the Royal Prussian Academy of Sciences (Berlin)*, p. 142 (in German).

Eisenstein, D. et al. (2005). Detection of the Baryon Acoustic Peak in the Large-Scale Correlation Function of SDSS Luminous Red Galaxies. *Astrophys. J.*, 633, pp. 560–74.

El-Ad, H. & Piran, T. (1997). Voids in the Large-Scale Structure. *Astrophys. J.*, 491, pp. 421–35.

El-Ad, H., Piran, T., & L. Da Costa, L. (1996). Automated Detection of Voids in Redshift Surveys. *Astrophys. J. Lett.*, 462, pp. L13–L16.

Falco, E., Kurtz, M., Geller, M., Huchra, J., Peters, J., Berlind, P., Mink, D., Tokarz, S., & Elwell, B. (1999). The Updated Zwicky Catalog (UZC). *Pub. Astron. Soc. Pacific*, 111, pp. 438–52.

Ferris, T. (1983). *The Red Limit* (New York, NY: Quill).

Ferris, T. (1989). *Coming of Age in the Milky Way* (New York, NY: Anchor Books, Doubleday).

Frenk, C., White, S., & Davis, M. (1983). Non-Linear Evolution of Large-Scale Structure in the Universe. *Astrophys. J.*, 271, pp. 417–30.

Friedmann, A. (1922). On the Curvature of Space. *General Relativity and Gravitation*, 31, p. 12, 1999 (English translation).

Fujimoto, M. (1983). Dynamics of Ellipsoidal Voids of Matter in an Expanding Universe. *Pub. Astron. Soc. Japan*, 35, pp.159–71.

Gamow, G. (1948a). Origin of the Elements and the Separation of Galaxies. *Phys. Rev.*, 74, 505–6.

Gamow, G. (1948b). The Evolution of the Universe. Nature, 162, pp. 680–2.

Geller, M. (1974). Bright Galaxies in Rich Clusters: A Statistical Model for Magnitude Distributions. Ph.D. thesis, Princeton University.

Geller, M. (1991) in *Realm of the Universe Fifth Edition, 1994 Version*. Abell, G., Morrison, D. & Wolff, S. (Fort Worth, TX: Saunders College Publishing),

Geller, M. & Huchra, J. (1989). Mapping the Universe. *Science*, 246, pp. 897–903.

Gibbons, G. & Hawking, S. (1977). Cosmological Event Horizons, Thermodynamics, and Particle Creation. *Phys. Rev.*, D15, pp. 2738–51.

Gingerich, O. (1990). Through Rugged Ways to Galaxies. *J. Hist. Astron.*, 21, 77–88.

Gingerich, O. (1999). Shapley, Hubble and Cosmology. In Edwin Hubble Centennial Symposium, ed. R. Kron, *A.S.P. Conference Series*, 10, pp. 19–21.

Ginzburg, V. (1994). Obituary: Yakov Borissovich Zel'Dovich, 8 March 1914 – 2 December 1987. Biographical Memoirs of Fellows of the Royal Society, 40, pp. 430–41.

Giovanelli, R., Haynes, M., & Chincarini, G. (1986). Morphological Segregation in the Pisces-Perseus Supercluster. *Astrophys. J.*, 300, pp. 77–92. Oort (1983) references this paper as a "preprint" dated (1981).

Gott, J. III, (2016). *The Cosmic Web* (Princeton, NJ: Princeton University Press).

Gott, J. III, Juric, M., Schlegel, D., Hoyle, F., Vogeley, M., Tegmark, M., Bahcall, N., & Brinkman, J. (2005). A Map of the Universe. *Astrophys. J.*, 624, 463–84.

Gott, J. III, Melott, A., & Dickinson, M. (1986). The Sponge-like Topology of Large-Scale Structure in the Universe. *Astrophys. J.*, 306, pp. 341–57.

Gott, J. III, Weinberg, D., & Melott, A. (1987). A Quantitative Approach to the Topology of Large-Scale Structure. *Astrophys. J.*, 319, pp. 1–8.

Graham, J., Wade, C., & Price, R. (1994). Bart J. Bok. *Natl. Acad., Sci.., Biographical Memoirs*, 64, pp. 72–97.

Gregory, S. (1975). Redshifts and Morphology of Galaxies in the Coma Cluster. *Astrophys. J.*, 199, pp. 1–9.

Gregory, S. & Thompson, L. (1978). The Coma/A1376 Supercluster and Its Environs. *Astrophys. J.*, 222, pp. 784–99.

Gregory, S. & Thompson, L. (1982). Superclusters and Voids in the Distribution of Galaxies. *Sci. Amer.*, 246, No. 3, pp. 106–14.

Gregory, S. & Thompson, L. (1984). The A2197 and A2199 Galaxy Clusters. *Astrophys. J.*, 286, pp. 422–36.

Gregory, S., Thompson, L., & Tifft, W. (1979). The Perseus/Pisces Supercluster. *Bull. Amer. Astron. Soc.*, 10, p.622.

Gregory, S., Thompson, L., & Tifft, W. (1981). The Perseus Supercluster. *Astrophys. J.*, 243, pp. 411–26.

Greenstein, J. (1974). Remembering Zwicky. *Science and Engineering Newsletter*. (Pasadena, CA: Caltech Library).

Grogin, N. & Geller, M. (1999). An Imaging and Spectroscopic Survey of Galaxies within Prominent Nearby Voids. I. The Sample and Luminosity Distribution" *Astron. J.*, 118, pp. 2561–80.

Grogin, N. & Geller, M. (2000). An Imaging and Spectroscopic Survey of Galaxies within Prominent Nearby Voids. II. Morphologies, Star Formation, and Faint Companions. *Astron. J.*, 119, pp. 32–43.

Guthrie, B. & Napier, W. (1991). Evidence for Redshift Periodicity in Nearby Field Galaxies. *Mon. Not. Royal Astron. Soc.*, 253, pp. 533–44.

Hamaus, N., Pisani, A., Sutter, P., Lavaux, G., Escoffier, S., Wandelt, B., & Weller, J. (2016). Constraints on Cosmology and Gravity from the Dynamics of Voids. *Phys. Rev. Lett.*, 117, 091302.

Harrison, E. (1970). Fluctuations at the Threshold of Classical Cosmology. *Phys. Rev. D*, 1, pp. 2726–30.

Hauser, M. & Peebles, P. (1973). Statistical Analysis of Catalogs of Extragalactic Objects: II. The Abell Catalog of Rich Clusters. *Astrophys. J.*, 185, pp. 757–85.

Hausman, M., Olson, D., & Roth, B. (1983). The Evolution of Voids in the Expanding Universe. *Astrophys. J.*, 270, pp. 351–9.

Herschel, W. (1784). Account of Some Observations Tending to Investigate the Construction of the Heavens. *Phil. Trans. Royal Soc. of London*, 74, pp. 437–51.

Herschel, W. (1811). Astronomical Observations Relating to the Construction of the Heavens, Arranged for the Purpose of a Critical Examination, the Result of

Which Appears to Throw Some New Light Upon the Organization of the Celestial Bodies. *Phil. Trans. Royal Soc. of London*, 101, pp. 437–51.

Hoffleit, D. (1992). *J. Amer. Assoc. of Variable Star Observers*, 21, pp. 151–6.

Hoffman, G., Salpeter, E., & Wasserman, I. (1983). Spherical Simulations of Holes and Honeycombs in Friedmann Universe. *Astrophys. J.*, 268, pp. 527–39.

Hoffman, Y. & Shaham, J. (1982). On the Origin of the Voids in the Galaxy Distribution. *Astrophys. J. Lett.*, 262, pp. L23–L26.

Holmberg, E. (1937). A Study of Double and Multiple Galaxies Together with Inquires into Some Metagalactic Problems with an Appendix Containing a Catalogue of 827 Double and Multiple Galaxies. *Medd. Lund Obs.*, 6, pp. 3–173.

Hoscheit, B. & Barger, A. (2018). The KBC Void: Consistency with Supernovae Type Ia and the Kinematic SZ Effect in a Lambda LTB Model. *Astrophys. J.*, 854, 46, 9 pp.

Hoyle, F. (1948). A New Model for the Expanding Universe. *Mon. Not. Royal Astron. Soc.*, 108, pp. 372–82.

Hoyle, F. & Vogeley, M. (2002). Voids in the Point Source Catalogue and the Updated Zwicky Catalog. *Astrophys. J.*, 566, pp. 641–51.

Hoyle, F. & Vogeley, M. (2004). Voids in the Two-Degree Field Galaxy Redshift Survey. *Astrophys. J.*, 607, pp. 751–64.

Hoyle, F. & Tayler, R. (1964). The Mystery of the Cosmic Helium Abundance. *Nature*, 203, pp. 1108–10.

Hoyt, B. (1980). Vesto Melvin Slipher. *Natl. Acad, Sci..*, *Biographical Memoirs*, 52, pp. 410–49.

Hubble, E. (1934). The Distribution of Extra-Galactic Nebuae. *Astrophys. J.*, 79, pp. 8–76.

Hubble, E. (1936a). Effects of Red Shifts on the Distribution of Nebulae. *Astrophys. J.*, 84, pp. 517–54.

Hubble, E. (1936b). *The Realm of the Nebulae* (New Haven, CT: Yale University Press).

Hubble, E. & Humason, M. (1931). The Velocity-Distance Relation among Extra-Galactic Nebulae. *Astrophys. J.*, 74, pp. 43–79.

Huchra, J. (2002). Interview of John Huchra by Patrick McCray on 15 February 2002, Niels Bohr Library & Archives, American Institute of Physics, College Park, MD USA, Aip.org/history-programs/niels-bohr-library/oral-histories/31280-2

Huchra, J. & Thuan, T. (1977). Isolated Galaxies. *Astrophys. J.*, 216, pp. 694–7.

Icke, V. (1973). Formation of Galaxies inside Clusters. *Astron. & Astrophys.*, 27, pp. 1–21.

Icke, V. (1984). Voids and Filaments. *Mon. Not. Royal Astron. Soc.*, 206, pp. P1–P3.

Jeans, J. (1919). The Present Position of the Nebular Hypothesis. *Popular Astronomy*, 27, pp. 339–48.

Joeveer, M., Einasto, J., & Tago, E. (1978). Spatial Distribution of Galaxies and Clusters of Galaxies in the Southern Galactic Hemisphere. *Mon. Not. Royal Astron. Soc.*, 185, pp. 357–70.

Joeveer, M. & Einasto, J. (1978). Has the Universe the Cell Structure? In The Large Scale Structure of the Universe, IAU Symposium No. 79, eds. M. Longair & J. Einasto, pp. 241–51.

Kaiser, N. (1984). On the Spatial Correlations of Abell Clusters. *Astrophys. J. Lett.*, 284, pp. L9–L12.

Kauffmann, G. & Fairall, A. (1991). Voids in the Distribution of Galaxies: An Assessment of their Significance and Derivation of a Void Spectrum. *Mon. Not. Royal Astron. Soc.*, 248, pp. 313–24.

Keenan, R., Barger, A., & Cowie, L. 2013. Evidence for a ~300 Megaparsec Scale Under-Density in the Local Galaxy Distribution. *Astrophys. J.*, 775, pp. 62–77.

Kenworthy, D., Scolnic, D., & Riess, A. (2019). The Local Perspective on the Hubble Tension: Local Structure Does Not Impact Measurement of the Hubble Constant. *Astrophys. J.*, 875, 145, 10pp.

Kirshner, R, Oemler, A. Jr., & Schechter, P. (1978). A Study of Field Galaxies. I. Redshifts and Photometry of a Complete Sample of Galaxies. *Astron. J.*, 83, pp. 1549–63.

Kirshner, R, Oemler, A. Jr., & Schechter, P. (1979). A Study of Field Galaxies. II. The Luminosity and Space Distribution of Galaxies. *Astron. J.*, 84, pp. 951–7.

Kirshner, R, Oemler, A. Jr., Schechter, P. & Shectman,S. (1981). A Million Cubic Megaparsec Void in Bootes. *Astrophys. J. Lett.*, 248, L57–L60.

Kirshner, R, Oemler, A. Jr., Schechter, P. & Shectman. S (1982). The Big Blank – Void in Space. *Sci. Amer.*, 246, No.2, pp. 75–83.

Kirshner, R., Oemler, A. Jr., Schechter, P., & Shectman, S. (1987). A Survey of the Bootes Void. *Astrophys. J.*, 314, pp. 493–506.

Klypin, A. & Shandarin, S. (1983). Three-Dimensional Numerical Model of the Formation of Large-Scale Structure in the Universe. *Mon. Not. Royal Astron. Soc.*, 204, pp. 891–907.

Koo, D., Kron, R., & Szalay, A. (1987). Deep redshift surveys of large-scale structure. Proceedings of the 13th Texas Symposium on Relativistic Astrophysics (Singapore and Teaneck, NJ: World Scientific Publishing Company) pp. 284–5.

Kreisch, C., Pisani, A., Carbone, C., Liu, J., Hawken, A., Massara, E., Spergel, D., & Wandelt, B. (2019). Massive Neutrinos Leave Fingerprints on Cosmic Voids. *Mon. Not. Royal Astron. Soc.*, 488, pp. 4413–26.

de Lapparent, V., Geller, M., & Huchra, J. (1986). A Slice of the Universe. *Astrophys. J. Lett.*, 302, pp. L1–L5.

Lavaux, G. & Wandelt, B. (2010). Precision Cosmology with Voids: Definition, Methods, Dynamics. *Mon. Not. Royal Astron. Soc.*, 403, pp. 1392–1408.

Lavaux, G. . Wandelt, B. (2012). Precision Cosmology with Stacked Voids. *Astrophys. J.*, 754, pp. 109–23.

Leavitt, H. (1908). 1777 Variables in the Magellanic Clouds. *Ann. Harvard College Obs.*, 60, No. 4, pp. 87–108.

Leavitt, H. & Pickering, E. (1912). Periods of 25 Variable Stars in the Small Magellanic Cloud. *Harvard College Obs.*, Circ. 173, pp. 1–3.

Lee, J. & Park, D. (2009). Constraining the Dark Energy Equation of State with Cosmic Voids. *Astrophys. J. Lett.*, 696, pp. L10–L12.

Lemaître, G. (1931a). A Homogeneous Universe of Constant Mass and Increasing Radius Accounting for the Radial Velocity of Extra-galactic Nebulae. *Mon. Not. Royal Astron. Soc.*, 91, 483–90.

Lemaître, G. (1931b). The Expanding Universe. *Mon. Not. Royal Astron. Soc.*, 91, 490–501.

Lightman, A. & Brawer, R. (1990). *Origins: The Lives and Worlds of Modern Cosmologists* (Cambridge, MA: Harvard University Press).

Limber, D. (1953). The Analysis of the Counts of the Extragalactic Nebulae in Terms of a Fluctuating Density Field. I. *Astrophys. J.*, 117, pp. 134–44.

Limber, D. (1954). The Analysis of the Counts of the Extragalactic Nebulae in Terms of a Fluctuating Density Field. II. *Astrophys. J.*, 119, pp. 655–81.

Limber, D. (1957). The Analysis of the Counts of the Extragalactic Nebulae in Terms of a Fluctuating Density Field. III. *Astrophys. J.*, 125, pp. 9–41.

Lin, C., Mestel, L. & Shu, F. (1965). The Gravitational Collapse of a Uniform Spheroid. *Astrophys. J.*, 142, pp. 1431–46.

Longair, M. (2006). *The Cosmic Century: A History of Astrophysics and Cosmology* (Cambridge, UK: Cambridge University Press).

Luud, L. Ruusalepp, E., & Kaasik, A. (1978). Anomalous Line Profiles for the Fe(42) Multiplet in the Spectrum of Deneb. *Sov. Astron. Lett.*, 4, pp. 151–2.

Lynden-Bell, D. (1964). On Large-Scale Instabilities during Gravitational Collapse and the Evolution of Shrinking Maclaurin Spheroids. *Astrophys. J.*, 139, pp. 1195–1216.

Lyubimov, V., Novikov, E., Nozik, V., Tretyakov E., & Kosik, V. (1980). An Estimate of the Electron Neutrino Mass from the Beta-Spectrum of Tritium in the Valine Molecule. *Physics Letters*, 138, pp. 30–56.

Madsen, C. (2013). Retirement of Massimo Tarenghi. *European Southern Obs. Messenger*, 153, pp. 39–41.

Matsuda, T. & Shima, E. (1984). Topology of Supercluster-Void Structure. *Progress of Theoretical Physics*, 71, pp. 855–8.

Mayall, N. (1951). Comparison of Rotational Motions Observed in the Spirals M31 and M33 and The Galaxy. *Pub. Obs. Michigan*, No. 10, p.19.

Mayall, N. (1960). Advantages of Electronic Photography for Extragalactic Spectroscopy. *Ann. Astrophys.*, 23, pp. 344–59.

Melott, A. (1983). Massive Neutrinos in Large-Scale Gravitational Clustering. *Astrophys. J.*, 264, pp. 59–86.

Melott, A. (1993). Galaxy Clustering: Why Peebles and Zeldovich Were Both Right. *Comments on Astrophys.*, 16, pp. 321–30.

Melott, A., Einasto, J., Saar, E., Suisalu, I., Klypin, A., & Shandarin, S. (1983). Cluster Analysis of the Non-Linear Evolution of Large-Scale Structure in an Axion / Gravitino / Photino Dominated Universe. *Phys. Rev. Letters*, 51, pp. 935–8.

Milne, E. (1935). *Relativity, Gravitation and World Structure*. (Oxford, UK: Oxford University Press).

Micheletti, D. ; 48 other authors (2014). The VIMOS Public Extragalactic Redshift Survey: Searching for Cosmic Voids. *Astron. & Astrophys.*, vol. 570, A106.

Moffat, J. & Tatarski, D. (1995). Cosmological Observations of a Local Void. *Astrophys. J.*, 453, pp. 17–24.

Moody, J., Kirshner, R., MacAlpine, G., & Gregory, S. (1987). Emission-Line Galaxies in the Bootes Void. *Astrophys. J. Lett.*, 314, pp. L33-L37.

Moss, A., Zibin, J., & Scott, D. (2011). Precision Cosmology Defeats Void Models for Acceleration. *Phys. Rev. D.*, 83, 103515.

Muller, V., Arbabi-Bidgoli,S., Einasto, J., & Tucker, D. (2000). Voids in the Las Campanas Redshift Survey versus Cold Dark Matter Models. *Mon. Not. Royal Astron. Soc.*, 318, pp. 280–8.

Nadathur, S., Carter, P., Percival, W., Winther, H., & Bautista, J. (2019). Beyond BAO: Improving Cosmological Constraints from BOSS with Measurement of the Void-Galaxy Cross-Correlation. *Phys. Rev. D*, 100, 023504.

Neyman, J., & Scott, E., (1952). A Theory of the Spatial Distribution of Galaxies. *Astrophys. J.*, 116, pp. 144–63.

Neyman, J, Scott, E., & Shane, C. (1953). On the Spatial Distribution of Galaxies: a Specific Model. *Astrophys. J.*, 117, pp. 92–133.

Neyrinck, M. (2008). ZOBOV: A Parameter-Free Void-Finding Algorithm. *Mon. Not. Royal Astron. Soc.*, 386, pp. 2101–9.

Oort, J. (1983). Superclusters. *Ann. Rev. Astron. & Astrophys.*, 21, pp. 373–428. (Palo Alto, CA: Annual Reviews).

Opik, E. (1922). An Estimate of the Distance of the Andromeda Nebula. *Astrophys. J.*, 55, pp. 406–10.

Osterbrock, D. (1993). *Pauper & Prince – Ritchey, Hale, and Big American Telescopes* (Tucson, AZ: University of Arizona Press).

Ostriker, J. & Cowie, L. (1981). Galaxy Formation in an Intergalactic Medium Dominated by Explosions. *Astrophys. J. Lett.*, 243, pp. 127–31.

Padilla, N., Ceccarelli, L., & Lambas, D. (2005). Spatial and Dynamical Properties of Voids in a Lambda Cold Dark Matter Universe. *Mon. Not. Royal Astron. Soc.*, 363, pp.977–90.

Pais, A. (1982). *Subtle is the Lord . . . The Science and the Life of Albert Einstein* (Oxford, UK: Oxford University Press).

Pan, D., Vogeley, M., Hoyle, F., Choi, Y., & Park, C. (2012). Cosmic Voids in Sloan Digital Sky Survey Data Release 7. *Mon. Not. Royal Astron. Soc.*, 421, pp.926–34.

Paz, D., Lares, M., Ceccarelli, L., Padilla, N., & Lambas, D. (2013). Clues on Void Evolution II. Measuring Density and Velocity Profiles on SDSS Galaxy Redshift Space Distortions. *Mon. Not. Royal Astron. Soc.*, 436, pp.3480–91.

Peacock, J. (2003). Large Scale Structure and Matter in the Universe. *Roy. Soc. London Trans. Ser. A*, 361, pp.2479–95.

Peebles, P. (1970). Structure of the Coma Cluster of Galaxies. *Astron. J.*, 75, pp.13–20.

Peebles, P. (1980). *The Large Scale Structure of the Universe* (Princeton, NJ: Princeton University Press).

Peebles, P. (1982a). The Peculiar Velocity around Holes in the Galaxy Distribution. *Astrophys. J.*, 257, pp. 438–41.

Peebles, P. (1982b). Large-Scale Background Temperature and Mass Fluctuations Due to Scale-Invariant Primeval Perturbations. *Astrophys. J.*, 263, pp. L1–L5.

Peebles, P. (1983). Hierarchical Clustering. in Clusters and Groups of Galaxies, eds. F. Mardirossian, G. Giuricin, & M. Mezetti, *Astrophys. & Space Sci. Library*, 111, pp. 405–14 (Dordrecht, Netherlands: D. Reidel).

Peebles, P. (1984). Interview of Jim Peebles by Martin Harwit on 1984 September 27, Niels Bohr Library & Archives, American Institute of Physics, College Park,MD USA, www.aip.org/history-programs/niels-bohr-library/oral-histories/33957

Peebles, P. (1988). Interview of Jim Peebles by Alan Lightman on 1988 January 19, Niels Bohr Library & Archives,American Institute of Physics, College Park,MD USA,www.aip.org/history-programs/niels-bohr-library/oral-histories/4814

Peebles, P. (1993). *Principles of Physical Cosmology* (Princeton, NJ: Princeton University Press).

Peebles, P. (2001). The Void Phenomenon. *Astrophys. J.*, 557, pp. 495–504.

Peebles, P. & Dicke, R. (1968). Origin of the Globular Clusters. *Astrophys. J.*, 154, pp. 891–908.

Peebles, P. & Yu, J. (1970). Primeval Adiabatic Perturbation in an Expanding Universe. *Astrophys. J.*, 162, pp.815–36.

Penzias, A. & Wilson, R. (1965). A Measurement of Excess Antenna Temperature at 4080 MHz. *Astrophys. J.*, 142, pp. 419–21.

Pisani, A., Massara, E., Spergel, D.; 26 co-authors. (2019). Cosmic Voids: A Novel Probe to Shed Light on our Universe. Astro2020 Science White Paper. *Bull. Amer. Astron. Soc.*, 51c, 40p.

Platen, E., Van de Weygaert, R., & Jones, B. (2007). A Cosmic Watershed: the WVF Void Detection Technique. *Mon. Not. Royal Astron. Soc.*, 380, pp.551–70.

Pomarede, D., Hoffman, Y., Courtois, H., & Tully, R. (2017). The Cosmic V-Web. *Astrophys. J.*, 845, pp.55–64.

Postman, M., Geller, M., & Huchra, J. (1986). The Cluster-Cluster Correlation Function. *Astron. J.*, 91, pp.1267–73.

Pranav, P. Edelsbrunner, H., Van de Weygaert, R., Vegter, G., Kerber, M., Jones, B., & Wintraecken, M. 2016. The Topology of the Cosmic Web in Terms of Persistent Betti Numbers. *Mon. Not. Royal Astron. Soc.*, 465, pp.4281–310.

Press, W. & Schechter, P. (1974). Formation of Galaxies and Clusters of Galaxies by Self-Similar Gravitational Condensation. *Astrophys. J.*, 187, pp.425–38.

Rees, M. (1977). Cosmology and Galaxy Formation. in Evolution of Galaxies and Stellar Populations. ed. B. Tinsley, and R. Larson. (New Haven, CT: Yale Univ. Obs.), pp.339–68.

Rees, M. & Ostriker, J. (1977). Cooling, Dynamics and Fragmentation of Massive Gas Clouds: Clues to the Masses and Radii of Galaxies and Clusters. *Mon. Not. Royal Astron. Soc.*, 179, pp.541–59.

Refsdal, S. (1964). The Gravitational Lens Effect. *Mon. Not. Royal Astron. Soc.*, 128, 295–306.

Ricciardelli, E., Cava, A., Varela, J., & Quilis, V. (2014). The Star Formation Activity in Cosmic Voids. *Mon. Not. Royal Astron. Soc.*, 445, pp.4045–54.

Riess, A., Macri, L., Casertano, S., Lampeid, H., Ferguson, H., Filippenko, A., Jha, S., Li, W., Chornock, R., & Silverman, J. (2011). A 3% Solution: Determination of the Hubble Constant with the Hubble Space Telescope and Wide Field Camera Three. *Astrophys. J.*, 730, 119, 18pp.

Rizzi, L., Tully, R., Shaya, E., Kourkchi, E., & Karachentsev, I. (2017). Draining the Local Void. *Astrophys. J.*, 835, pp.78–85.

Roberts, M. (1966). A High-Resolution 21-cm Hydrogen-Line Survey of the Andromeda Nebula. *Astrophys. J.*, 144, pp.639–56.

Rojas, R., Vogeley, M., Hoyle, F., & Brinkmann, J. (2004). Photometric Properties of Void Galaxies in the Sloan Digital Sky Survey). *Astrophys. J.*, 617, pp.50–63.

Roll, P. & Wilkinson, D. (1966). Cosmic Background Radiation at 3.2 cm – Support for Cosmic Black-Body Radiation. *Phys. Rev. Lett.*, 16, pp. 405–7.

Rood, H. (1988a). Voids. *Ann. Rev. Astron. & Astrophys.*, 26, pp.245–94.

Rood, H. (1988b). Supplementary Topics on Voids. *Pub. Astron. Soc. Pacific*, 100, pp. 1071–5.

Rubin, V. (1951). Differential Rotation of the Inner Metagalaxy. *Astron. J.*, 56, pp. 47–8.

Rubin, V. (1954). Fluctuation of the Space Distribution of the Galaxies. *Pub. National Acad. Sci.*, 40, 541–9.

Rubin, V. (1989). The Local Supercluster. Ch.16. in *Gerard and Antoinette de Vaucouleurs: A Life for Astronomy*, eds. M. Capaccioli & H. Corwrin, Jr. (Singapore: World Scientific Publishing Co.).

Rubin, V. & Ford, W., Jr. (1970). Rotation of the Andromeda Nebula from a Spectroscopic Survey of Emission Regions. *Astrophys.J.*, 159, pp.379–403.

Rubin, V., Ford, W., Jr., & Thonard, N. (1978). Extended Rotation Curves of High-Luminosity Spiral Galaxies. IV. Systematic Dynamical Properties, Sa through Sc. *Astrophys. J. Lett.*, 225, pp.L107–L111.

Ryden, B. (1995). Measuring q_o from the Distortion of Voids in Redshift Space. *Astrophys. J.*, 452, pp.25–32.

Sahlen, M., Zubeldia, I., & Silk, J. (2016). Cluster-Void Degeneracy Breaking: Dark Energy, Planck, and the Largest Cluster & Void. *Astrophys. J. Lett.*, 820, L7–L12.

Sandage, A. (1961). The Ability of the 200-inch Telescope to Discriminate between Selected World Models. *Astrophys. J.*, 133, pp.355–92.

Sandage, A. (1987) . Observational Cosmology 1920 – 1985: An Introduction to the Conference. in *I.A.U. Symposium No. 124, Observational Cosmology*, eds. A. Hewitt, G. Burbidge, & L. Fang. (Dordrecht, Netherlands: D. Reidel), pp.1–27.

Sandage, A. (1989). Edwin Hubble 1889-1953. *J. Royal Astron. Soc. Canada*, 83, pp.351–62.

Sandage, A., Tammann, G., & Hardy, E. (1972). Limits on the Local Deviation of the Universe from a Homogeneous Model. *Astrophys. J.*, 172, pp.253–63.

Sanduleak, N. & Pesch, P. (1987). The Case Low-Dispersion Northern Sky Survey. IV - Galaxies in the Bootes Void Region. *Astrophys. J. Suppl.*, 63, pp.809–19.

Sarkar, S., Pandey, B., & Khatri, R. (2019). Testing Isotropy in the Universe using Photometric and Spectroscopic Data from the SDSS. *Mon. Not. Royal Astron. Soc.*, vol. 483, pp.2453–64.

Saunders, W., Sutherland, W., Maddox, S., Keeble, O., Oliver, S., Rowan-Robinson, M., McMahon, R., Efstathiou, G., Tadros, H., White, S., Frenk, C., Carraminana, A., & Hawkins, M. (2000). The PSCz Catalogue. *Mon. Not. Royal Astron. Soc.*, 317, pp. 55–63.

Scott, E., Shane, C., & Swanson, M. (1954). Comparison of the Synthetic and Actual Distribution of Galaxies on a Photographic Plate. *Astrophys. J.*, 119, 91–112.

Shaikh, S., Mukherjee, S., Das, S., Wandelt, B., & Souradeep, T. (2019). Joint Bayesian Analysis of Large angular scale CMB Temperature Anomalies. *J. Cosmol. Astroparticle Phys.*, 08, 007.

Shane, C. (1970). Distribution of Galaxies. In *Galaxies and the Universe*, ed. A. Sandage, M. Sandage, & J. Kristian (Chicago, IL: University of Chicago Press), pp.647–63.

Shane, C. & Wirtanen, C. (1948). The Distribution of Extragalactic Nebulae. *Lick Obs. Bulletin*, 20, pp.91–110.

Shane, C. & Wirtanen, C. (1954). The Distribution of Extragalactic Nebulae. *Astron. J.*, 59, pp. 285–304.

Shapley, H. (1919). Studies Based on the Colors and Magnitudes in Stellar Clusters. Twelfth Paper: Remarks on the Arrangement of the Sidereal Universe. *Astrophys. J.*, 49, pp.311–33.

Shapley, H. (1930a). The Super-Galaxy Hypothesis. *Harvard College Obs. Circ.*, 350, pp.1–7.

Shapley, H. (1930b). Note on a Remote Cloud of Galaxies in Centaurus. *Harvard College Obs. Bulletin*, 874, pp.9–12.

Shapley, H. (1934). A First Search for a Metagalactic Gradient. *Harvard College Obs. Bulletin*, 894, pp. 5–13.

Shapley, H. (1938). A Metagalactic Density Gradient. *Publ. Natl. Acad. Sci.*, 24, pp.282–7.

Shapley, H. & Ames, A. (1932a). A Survey of the External Galaxies Brighter that the Thirteenth Magnitude. *Ann. Astron. Observat. Harvard College*, 88, No. 2, pp.43–75. (Shapley-Ames Catalogue).

Shapley, H. & Ames, A. (1932b). Photometric Survey of the Nearer Extragalactic Nebulae. *Harvard College Observat. Bull.*, No. 887, pp.1–6.

Shaya, E., Tully, R., Hoffman, Y., & Pomarede, D. (2017). Action Dynamics of the Local Supercluster. *Astrophys. J.*, 850, pp.207–22.

Sheth, R. & Van de Weygaert, R. (2004). A Hierarchy of Voids: Much Ado about Nothing. *Mon. Not. Royal Astron. Soc.*, 350, pp.517–38.

Slipher, V. (1917). Nebulae. *Proc. American Philosophical Society*, 56, pp.403–9.

Smith, R. (1982). *The Expanding Universe: Astronomy's "Great Debate" 1900–1931* (Cambridge, UK: Cambridge University Press).

Soneira, R. & Peebles, P. (1978). A Computer Model Universe – Simulation of the Nature of the Galaxy Distribution in the Lick Catalog. *Astron. J.*, 83, pp.845–60.

Steinicke, W. (2010). *Nebulae and Star Clusters: From Herschel to Dreyer's New General Catalogue* (Cambridge University Press: Cambridge, England).

Sunyaev, R. & Zeldovich, Ya. (1972). Formation of Clusters of Galaxies; Protocluster Fragmentation and Intergalactic Gas Heating. *Astron. & Astrophys.*, 20, pp.189–200.

Sutter, P., Lavaux, G., Wandelt, B., & Weinberg, D. (2012). A Public Void Catalog from the SDSS DR7 Galaxy Redshift Surveys Based on the Watershed Transform. *Astrophys. J.*, 761, pp.44–56.

Szalay, A. & Marx, G. (1976). Neutrino Rest Mass from Cosmology. *Astron. & Astrophys.*, 49, pp.437–41.

Tarenghi, M., Tifft, W., Chincarini, G., Rood, H., & Thompson, L. (1978). The Structure of the Hercules Supercluster. In *The Large Scale Structure of the Universe, IAU Symposium No. 79*, eds. M. Longair & J. Einasto. (Dordrecht, Netherlands: D. Reidel Publishing Co.), pp. 263–5.

Tarenghi, M., Tifft, W., Chincarini, G., Rood, H., & Thompson, L. (1979). The Hercules Supercluster. I. Basic Data. *Astrophys. J.*, 234, pp.793–801.

Tarenghi, M., Tifft, W., Chincarini, G., Rood, H., & Thompson, L. (1979) . The Hercules Supercluster. II. Analysis. *Astrophys. J.*, 235, pp.724–42.

Thompson, L. (1976). Angular Momentum Properties of Galaxies in Rich Clusters. *Astrophys. J.*, 209, pp. 22–34.

Thompson, L. (1977). Possible Ring Galaxies near Rich Clusters. *Astrophys. J.*, 211, pp.684–92.

Thompson, L. (1983). Markarian Galaxies and Voids in the Galaxy Distribution. *Astrophys. J. Lett.*, 266, pp.446–50.

Thompson, L. (2013). Slipher and the Development of the Nebular Spectrograph. In *Origins of the Expanding Universe: 1912–1932*, eds. M. Way & D Hunter. Astron. Soc. Pacific. Conference Series, 471, pp.135–42.

Thompson, L. & Gardner, C. (1987). Experiments on Laser Guide Stars at Mauna Kea Observatory for Adaptive Imaging in Astronomy. Nature, 238, pp.229–31.

Thompson, L. & Gregory, S. (1978). Is the Coma Cluster a Zeldovich Disk? *Astrophys. J.*, 220, 809–13.

Tifft, W. (1972a). Two Dimensional Area Scanning with Image Dissectors. *Pub. Astron. Soc. Pacific*, 84, pp.137–44.

Tifft, W. (1972b). The Correlation of Redshift with Magnitude and Morphology in the Coma Cluster. *Astrophys. J.*, 175, pp.613–35.

Tifft, W. (1976). Discrete States of Redshift & Galaxy Dynamics. I. Internal Motions in Single Galaxies. *Astrophys. J.*, 206, pp.38–56.

Tifft, W. & Gregory, S. (1976). Direct Observations of the Large-Scale Distribution of Galaxies. *Astrophys. J.*, 205, pp.696–708.

Tifft, W & Gregory, S. (1978). Observations of the Large Scale Distribution of Galaxies. In *The Large Scale Structure of the Universe, IAU Symposium No. 79*, eds. M. Longair & J. Einasto. (Dordrecht, Netherlands: D. Reidel Publishing Co.), pp. 267–9.

Tinker, J. & Conroy, C. (2009). The Void Phenomenon Explained. *Astrophys. J.*, 691, pp.633–9.

Tolman, R. (1931). On the Problem of the Entropy of the Universe as a Whole. *Phys. Rev.*, 37, pp.1639–60.

Tolman, R. (1934). Effect of Inhomogeneity on Cosmological Models. *Proc. Natl. Acad., Sci.*, 20, pp.169–76.

Tremaine, S. & Gunn, J. (1979). Dynamical Role of Light Leptons in Cosmology. *Phys. Rev. Lett.*, 42, pp.407–10.

Trumpler, R. (1930). Absorption of Light in the Galactic System. *Pub. Astron. Soc. Pacific*, 42, pp.214–27.

Tully, R. (1982). The Local Supercluster. *Astrophys. J.*, 257, pp.389–422.

Tully, R. & Fisher, J. (1978a). Nearby Small Groups of Galaxies. In *The Large Scale Structure of the Universe, IAU Symposium No. 79*, eds. M. Longair & J. Einasto. (Dordrecht, Netherlands: D. Reidel Publishing Co.), pp.31–48.

Tully, R. & Fisher, J. (1978b). A Tour of the Local Supercluster. In *The Large Scale Structure of the Universe, IAU Symposium No. 79*, eds. M. Longair & J. Einasto. (Dordrecht, Netherlands: D. Reidel Publishing Co.), pp. 214–16.

Tully, R. & Fisher, J. (1987). *Nearby Galaxies Atlas* (Cambridge, UK: Cambridge University Press).

Tully, R., Pomarede, D., Graziana, R., Courtois, H., Hoffman, Y., & Shaya, E. (2019). Cosmicflows-3: Cosmography of the Local Void. *Astrophys. J.*, 880, i.d.24, 14 pp.

Tully, R., Shaya, E., Karachentsev, I., Courtois, H., Kocevski, D., Ruzzi, L., & Peel, A. (2008). Our Peculiar Motion away from the Local Void. *Astrophys. J.*, 676, pp.184–205.

Yu, J. & Peebles, P. (1969). Superclusters of Galaxies. *Astrophys. J.*, 158, pp.103–13.

van de Weygaert, R., Kreckel, K., Platen E., ; 9 co-authors. (2011). The Void Galaxy Survey. *Astrophys. Space Sci. Proceedings*, 27, pp.17–25.

de Vaucouluers, G. (1953). Evidence for a Local Supergalaxy. *Astron. J.*, 58, pp.30–2.

de Vaucouleurs, G. (1956). The Distribution of Bright Galaxies and the Local Supergalaxy. *Vistas in Astronomy*, 2, 1584 (London, England: Pergamon Press), pp.1584–606.

de Vaucouleurs, G. (1970). The Case for a Hierarchical Cosmology. *Science*, 167, pp.1203–13.

de Vaucouleurs, G. (1971). The Large Scale Distribution of Galaxies and Clusters of Galaxies., *Pub. Astron. Soc. Pacific.*, 83, pp.113–43.

de Vaucouleurs, G. (1988). Interview of G. de Vaucouleurs by Alan Lightman on 1988 November 7, Niels Bohr Library & Archives, American Institute of Physics, College Park, MD USA, www.aip.org/history-programs/niels-bohr-library/oral-histories/33930

de Vaucouleurs, G. (1991). Interview of Gerard de Vaucouleurs by Ronald Doel on 1991 November 23, Niels Bohr Library & Archives, American Institute of Physics, College Park, MD USA, www.aip.org/history-programs/niels-bohr-library/oral-histories/31929-2

Vettolani, G., De Souza, R., Marano, B., & Chincarini, G. (1985). The Distribution of Voids. *Astron. Astrophys.*, 144, pp.506–13.

Vogeley, M. (1993). Statistical Measures of the Large-Scale Structure. Ph.D. thesis, Harvard University.

Vogeley, M., Geller, M., & Huchra, J. (1991). Void Statistics of the CfA Redshift Survey. *Astrophys. J.*, 382, pp.44–54.

Vorontsov-Velyaminov, B. (1987). *Extragalactic Astronomy: Revised and Expanded English Edition*. Translated from the Russian by R. Rodman. Edited by N. Sharp. Supplemented by D. Meloy Elmergreen. With contributions by H. Bushouse & H. Corwin (Chur, Switzerland: Harwood Academic Publishers), pp. 483–484.

Wagoner, R., Fowler, W., & Hoyle, F. (1967). On the Synthesis of Elements at Very High Temperature. *Astrophys. J.*, 148, pp. 3–49.

Weistrop, D. & R. Downes, R. 1988). Spectra of Galaxies in the Case Low-Dispersion Sky Survey in the Direction of the Bootes Void. *Astrophys. J.*, 331, pp.172–80 .

Wertz, J. (1970). Newtonian Hierarchical Cosmology. Ph.D. Thesis, Univ. of Texas.

Wertz, J. (1971). A Newtonian Big-Bang Hierarchical Cosmological Model. *Astrophys. J.*, 164, pp.227–36.

Whitbourn, J. & Shanks, T. (2016). The Galaxy Luminosity Function and the Local Hole. *Mon. Not. Royal Astron. Soc.*, 459, pp.496–507.

White, S. (1976). The Dynamics of Rich Clusters of Galaxies. *Mon. Not. Royal Astron. Soc.*, 177, pp.717–33.

White, S. (1979). The Hierarchy of Correlation Functions and Its Relation to Other Measures of Galaxy Clustering. *Mon. Not. Royal Astron. Soc.*, 186, pp. 145–54.

White, S. (2017). Reconstructing the Universe in a Computer: Physical Understanding in the Digital Age, arXiv:1806.06348.

White, S., Frenk, C., & Davis, M. (1983). Clustering in a Neutrino-Dominated Universe. *Astrophys. J.*, 274, L1–L5.

White, S., Frenk, C. & Davis, M., & Efstathiou, G. (1987). Clusters, Filaments, and Voids in a Universe Dominated by Cold Dark Matter. *Astrophys. J.*, 313, pp. 505–16. (WFDE).

White, S. & Rees, M., (1978). Core Condensation in Heavy Halos – A Two-Stage Theory for Galaxy Formation and Clustering. *Mon. Not. Royal Astron. Soc.*, 183, pp. 341–58.

Yo, H-J. (2017). Does the Universe Really Expand, or Does the Size of Matter Shrink Instead?. *J. Modern Phys.*, 8, pp.2077–86.

York, D., Adelman, J., Anderson, J. Jr., & 142 other coauthors. (2000). The Sloan Digital Sky Survey: Technical Summary. *Astron. J.*, 120, pp.1579–87.

Zeldovich, Y. (1970). Gravitational Instability: An Approximate Theory for Large Density Perturbations. *Astron. Astrophys.*, 5, pp. 84–9.

Zeldovich, Y. (1972). A Hypothesis, Unifying the Structure and the Entropy of the Universe. *Mon. Not. Royal Astron.*, 160, pp. 1P–3P.

Zeldovich, Y. (1983). Modern Cosmology. Highlights of Astronomy, v. 6, in *Proceedings of the 18th IAU General Assembly* . ed. R. West (Dordrecht: Netherlands: D. Reidel), pp.29–52.

Zeldovich, Y., Einasto, J., & Shandarin, S. (1982). Giant Voids in the Universe. *Nature*, 300, pp. 407–13.

Zeldovich, Y. & Sunyaev, R. (1980). Astrophysical Implications of the Neutrino Rest Mass. I. The Universe. *Astron. Lett. (Russian)*, 6, pp.451–56.

Zwicky, F. (1933). Die Rotverschiebung von Extragalaktischen Nebeln. *Helvetica Physics*, 6, pp.110–27.

Zwicky. F. (1937). On the Masses of Nebulae and of Clusters of Nebulae. *Astrophys. J.*, 86, pp.217–46.

Zwicky, F. (1957). *Morphological Astronomy* (Berlin, Germany: Springer-Verlag).

Zwicky, F. (1959). Multiple Galaxies. In *Encyclopedia of Physics*, 53, ed. S. Flugge (Berlin, Germany: Springer Verlag, 1959.), p. 396.

Zwicky, F., Herzog, E., & Wild, P. (1961). *Catalogue of Galaxies and of Clusters of Galaxies*. 1, p. 24 (Zurich, Switzerland: California Institute of Technology).

Index